IMAGES
of Aviation
NAVAL AIR STATION
ATLANTIC CITY

VX-3 FJ-3s, September 1954. The skies above Atlantic City have seen a spectrum of aircraft, from a Wright Flyer to Navy fighters and attack planes, New Jersey Air National Guard jet fighters, lifesaving Coast Guard helicopters, FAA test aircraft, and airliners. Here, Navy FJ-3 Furies overfly Bader Field with Atlantic City in the background. Convention Hall, now called Boardwalk Hall, is seen behind XC/92. (Tailhook Association.)

On the Cover: Air Development Squadron 3 of Naval Air Station Atlantic City, c. 1954. Tasked with developing and evaluating tactics and procedures for new naval aircraft, this squadron's complement of aircraft was always diverse. From the top are the following: a Lockheed TV-2 trainer; Douglas AD Skyraider; McDonnell F2H-2P Photo Banshee and F2H-2 Banshee; aluminum finished F2H-3 "Big Banjo" Banshee; Grumman F9F-5 Panther; and its swept-wing development, F9F-6 Cougar. (Tailhook Association.)

IMAGES of Aviation
Naval Air Station Atlantic City

Richard V. Porcelli

Copyright © 2012 by Richard V. Porcelli
ISBN 978-0-7385-7670-1

Published by Arcadia Publishing
Charleston, South Carolina

Printed in the United States of America

Library of Congress Control Number: 2011933502

For all general information, please contact Arcadia Publishing:
Telephone 843-853-2070
Fax 843-853-0044
E-mail sales@arcadiapublishing.com
For customer service and orders:
Toll-Free 1-888-313-2665

Visit us on the Internet at www.arcadiapublishing.com

*To Linda, my beacon who has guided me to follow new paths,
and . . . to all the men and women of our armed forces who
give so much of themselves for our freedom and security*

Contents

Acknowledgments		6
Introduction		7
1.	AC Fledglings and First "Air Port"	11
2.	Pine Barrens to Fighter Base	21
3.	Postwar Technical Developments and Korea	39
4.	Goodbye Navy . . . Hello FAA	89
5.	Guarding the State, Protecting the Nation	103
6.	The Spirit of Atlantic City	119

Acknowledgments

One of the greatest personal rewards in writing this book has been meeting so many people who shared my enthusiasm for this project and who provided me with so much assistance. Without their help, this project would have been impossible. My special thanks go to Stan Ciurczak of the FAA's William J. Hughes Technical Center, who provided support throughout my research; Heather Perez and Pat Rothenberg of the Atlantic City Free Public Library and June Sheridan of the Greate Egg Harbour Township Historical Society, who happily opened their files to me; Comdr. Doug Siegfried of the Tailhook Association, Andy Kondrach and Joan Legg of the Millville Army Air Field Museum, Sgts. Andrew Moseley and Matt Hecht of the 177th Fighter Wing of the New Jersey Air National Guard, and Lts. Augustino Albanese and Jay Kircher of USCG Air Station Atlantic City; Bette Epstein of the New Jersey State Archives, Jack Coyle of the National Naval Aviation Museum, Dick Atkins and Roger Stites of the Vought Aircraft Heritage Center, Patrick McGee, USAF (ret.), Robert Hanshew of the Naval History and Heritage Command, and Holly Reed, along with the great staff of the National Archives, who made gathering so many historic photographs an enjoyable task. I also thank Lt. Col. (ret.) David Haar and author Steve Ginter for their personal encouragement for this project, Diane Miller of the Atlantic Heritage Center, Sharon Gordon of the South Jersey Transportation Authority, Super Sabre aficionado Michael Benolkin for valuable information, Sgt. Jennifer Lindsey, USAF, for approval for historical file access, and my brothers Joe and Bob for sparking my lifelong passion for all things aeronautical.

Finally, I owe an unrepayable debt of gratitude to my wife, Linda, who has encouraged me to follow my dreams, who has provided me unwavering support in all things I have attempted for as long as we have known each other, and who has been an invaluable research assistant, proof reader, sounding board, counselor, organizer, and best friend throughout this book project.

Photograph credits:
ACFPL, Atlantic City Free Public Library
GEH, Greate Egg Harbour Historical Society
MAAM, Millville Army Airfield Museum
NA, National Archives
NH, Naval Historical Center
NJANG, New Jersey Air National Guard
NJSA, New Jersey State Archives
NNAM, National Naval Aviation Museum
Patrick McGee, USAF (ret.)
Stan Ciurczak, FAA William J. Hughes Technical Center
TH, Tailhook Association
USCG, US Coast Guard
Vought, Vought Archives

INTRODUCTION

When one thinks of Atlantic City, one pictures a seaside playground that went through some "lean years;" a gambling resort that does not quite rival the glitz and celebrity of Las Vegas; Burt Parks singing "There She Is" during a Miss America pageant; the city being the basis for Monopoly (perhaps, the most famous board game of all time); and the subject of a popular cable-television miniseries. Few would guess that Atlantic City has a rich aviation history, being instrumental in many technical developments as well as playing a key role in the defense of the nation.

Although the area's aviation history dates back to before World War I, the focus of this book is Naval Air Station Atlantic City (NASAC) as well as its transition after the Navy left in 1958 to ownership by the Federal Aviation Administration. Therefore, this book also looks at the FAA's William J. Hughes Technical Center activities, as well as those of its "tenants:" the Air National Guard, Coast Guard Air Station, Department of Homeland Security's Transportation Security Laboratory and Federal Air Marshal Training Center, and commercial airport.

A mere seven years after the Wright brothers' first manned flight of a heavier-than-air machine, the Aero Club of Atlantic City hosted the fourth air meet in history. From July 4 to 12, more than 100,000 spectators witnessed the aviation greats of the time give demonstrations and set records.

Immediately following the "Aero Show," Atlantic City's first airport, later named Bader Field, opened. It was the nation's first municipal airfield with facilities for both land and sea planes. In 1910, "aeronaut" (the name given the few fortunate "air travelers" of the day) Augustus Post coined the term "air port" to describe Atlantic City's prominent position in aviation development. In 1919, famed author Robert Woodhouse applied the term "air port" to Atlantic City's Bader Field, which, until its closure in 2006, was the longest continuously operating public airfield in the country. During the 1920s and 1930s, a number of airlines flew regularly scheduled flights to Philadelphia, New York City, Washington, DC, and other cities from Bader Field.

As commercial air travel grew, the Atlantic City government and business leaders recognized the need for a new, larger municipal airport. Atlantic City acquired about 5,000 acres of pine trees and marshland west of the city for the airport, a reservoir, and other future needs. Toward this end, in 1941, the city and Civil Aeronautics Administration, using Work Projects Administration funds, began construction of a new airport in Pomona, New Jersey. The start of war changed the plans for this site as the Navy added Atlantic City to its network of military airfields along the Eastern Seaboard.

The Navy reached an agreement with Atlantic City to lease the site of the proposed new municipal airport for 20 years, with provision for its return to the city six months after the end of hostilities. Thousands of civilian workers descended on the site to finish clearing the land, constructing hangars and buildings and laying four one-mile-long runways.

Naval Air Station Atlantic City (NASAC) was commissioned on April 24, 1943. Aircraft actually started to arrive earlier from air stations at Corpus Christi and Norfolk but had to be kept at Bader Field since the new air station's taxiways were not completed. From Bader, they had to be towed to the air station over city streets and roads.

The initial units commissioned on May 1, 1943, at NASAC were fighter and composite (scout and torpedo bomber) squadrons from Carrier Air Groups 31 and 25, mainly for training but also for patrolling the coastline. Also assigned was Carrier Air Service Unit 23 (CASU-23), whose role was to provide maintenance for aircraft assigned to the fleet.

After a brief period of training entire air groups, consisting of fighter, torpedo bomber, and scout/dive bomber aircraft for fleet aircraft carriers, the role was changed in August 1943 to become one of the Navy's premier Fighter Training Unit bases. South Pacific combat veteran and ace Lt. Stanley Vijtasa was assigned to organize this training unit. He staffed the unit with other seasoned combat veterans to teach the newly minted naval aviators high- and low-level gunnery, field carrier practice landings, catapult launches, dive and glide bombing as well as strafing and rocket attacks.

At a time when air-to-ground radio communications was in its infancy, the Fighter Training Unit also established a Combat Information Center (CIC), which was a training center where sailors were taught to control and coordinate the air defense of carriers using radar and radio communications—the basis of what would become the Air Traffic Control system. They established procedures, tested equipment, and trained radar operators, air combat controllers, and fighter pilots in the art of coordinated, radar-controlled defense of carrier task forces. For this purpose, the Navy took over the Brigantine Hotel and equipped it with radar, radio communications, and control rooms. In addition to training, this facility also assisted aviators caught in bad weather or just lost finding their way back to the air station.

NASAC also operated other facilities to support the training of fighter pilots. Outlaying landing fields were established at Woodbine and Ocean City as well as Bader Field. Bombing, rocket-firing, and strafing target ranges were set up at Coyle, Great Bay near Brigantine, Tuckahoe, Jeffers Landing, and Warren Grove—the latter range is still used by the Air National Guard. The air station also operated crash boats from Gardiner's Basin in Atlantic City for the rescue of aviators who had to abandon their aircraft over the sea.

By the end of the war, more than 50 fighter squadrons, flying Grumman Hellcats, and Vought Corsairs had passed through NASAC on the way to action in the Atlantic and Pacific theaters of operation. At its peak, aircraft onboard totaled 275. After the war, the pace of technical innovation and testing at NASAC did not slacken. Initially, VT-58, one of the last torpedo squadrons in the Navy, was commissioned at NASAC flying Grumman Avengers. It was reestablished as Attack Squadron 1 Lima (VA-1L) and joined by Fighting Squadron 1 Lima (VF-1L) flying Corsairs and Bearcats. In addition, the air station prepared Hellcat drones for monitoring the 1946 atomic bomb tests at Bikini Atoll.

In 1948, Atlantic City became the home of Air Development Squadron 3 (VX-3), which was formed by merging VA-1L and VF-1L. This large squadron developed and tested tactics for air warfare and night operations. It operated at NASAC until its decommissioning in 1958, a period of prolific development of exciting and challenging new naval aircraft.

With the 1952 commissioning of Composite Squadrons 4 (VC-4) and 33 (VC-33), NASAC become an All-Weather Master Jet Naval Air Station. VC-4 was tasked with providing the fleet night and all-weather fighter protection, as well as "special missions" (i.e., delivery of nuclear weapons). This large squadron employed multiple fighter types and provided aircraft detachments to fleet aircraft carriers. VC-33 played a similar role in providing all-weather and night attack capabilities. Being an all-weather and night flying center, the ground air control expertise developed during the war was expanded into a Fleet Electronic Training Unit for Ground-Controlled Intercept (GCI) and Ground-Controlled Approach (GCA) procedures.

VC-4 and VC-33 provided detachments for Korean War cruises, flying Corsairs and Douglas AD Skyraiders from the carriers *Boxer*, *Leyte*, and *Bon Homme Richard*. VC-4 also provided a detachment of Douglas F3D Skyknight night fighters on *Lake Champlain*, but for operational reasons, they were sent to a Korean base to fly alongside a Marine night-fighting unit; the combined unit was credited with 10 nighttime kills.

The Navy decided to decommission NASAC in 1958. Two years earlier, the infamous midair collision of two commercial airliners over the Grand Canyon shook the public's confidence in air travel. This spurred the Eisenhower administration to create the Airways Modernization Board (AMB), a temporary measure to coordinate avionics developments to aid air safety and to establish a technology center to develop equipment and procedures to meet the needs of safe and efficient air travel. The AMB selected the former NASAC, boasting long runways, radar facilities, and infrastructure, as the site of its National Aviation Facilities Experimental Center (NAFEC). Shortly thereafter, the AMB was superseded by the Federal Aviation Agency (FAA). NAFEC, now known as the William J. Hughes Technical Center, is the site of specialized facilities, including air traffic control simulators, a human factors laboratory, a full-scale fire test facility with a chemistry laboratory for analysis of combustion products, a diverse fleet of test aircraft, radar research and development laboratories, an aircraft structures test lab, and the National Airport Pavement Test Facility. Achievements include advances in automated air traffic control, understanding wake turbulence (potentially dangerous disturbance of air caused by a preceding aircraft), testing of the Automated Enroute Air Traffic Control system, development of the Visual Approach Slope Indicator for landing guidance, evaluation of Traffic Collision Avoidance System to avoid midair collisions, development of a frangible pavement for runway overrun safety, and control tower and controller console modernization. Today, it is the Center of Excellence, leading the switch from ground-based to satellite-based air traffic control, a system known as NextGen. It can honestly be said that every air traveler today benefits from safety and security advances made at Atlantic City.

With the FAA's acquisition of the former NASAC in Pomona and expansion to about 5,000 acres, it also gained a number of "tenant" organizations. The Navy's departure from Atlantic City coincided with the need of a new home for the New Jersey Air National Guard's 177th Fighter Wing/119th Fighter Squadron—a unit that can claim one of the oldest histories in the Air Force. The advent of high-performance jet aircraft and the growth in commercial air traffic made its base at Newark Airport untenable. The availability of a fully equipped jet base at Atlantic City presented an opportunity to the 177th FW that was "too good to refuse." The Jersey Devils have operated a number of jet fighter types over the years with both air defense and general-purpose assignments. The unit moved to Atlantic City with F-84F Thunderstreak fighter-bombers, which were subsequently replaced with F-100C Super Sabres, and F-105B Thunderchiefs. The unit was called to active duty twice, for the Berlin and Pueblo crises. The role was switched to air defense in 1972 when they converted to the F-106 Delta Dart. Today, the 177th Fighter Wing is flying 18 F-16C Fighting Falcons ("Vipers") from the ramp which once was the home of Navy Corsairs, Hellcats, and early jet fighters. They have provided invaluable protection of the Eastern Seaboard throughout the Cold War and then after the terrorist attacks of 9/11, participated in Operations Noble Eagle, Enduring Freedom, and Iraqi Freedom. A little known fact is the airport's 10,000-foot runway and the presence of the NJANG's well-trained emergency services resulted in Atlantic City being designated as a space shuttle emergency abort landing site.

Another important "tenant" is the US Coast Guard (USCG). In order to enhance efficiency and reduce operating costs, the Coast Guard closed two historic air stations, Brooklyn/Floyd Bennett Field and Cape May, and combined its operations at a new air station at Atlantic City. Air Station Atlantic City is the largest single-type air station in the Coast Guard. It flies 10 of the latest MH-65D helicopters, providing lifesaving and port protection services from Connecticut to Virginia, including a detachment to Reagan Airport for the safeguarding of the air space around the nation's capital.

Perhaps one of the most crucial recent developments is the growing presence of the Department of Homeland Security, including the Transportation Security Laboratory, tasked with research, development and validation of methods for detecting and mitigating the threat to air safety from explosives. Atlantic City is also home of a federal air marshal training center.

The FAA relinquished ownership of the commercial airfield, now known as Atlantic City International Airport, to the South Jersey Transportation Authority, which is focusing on developing commercial air service for Atlantic City as well as the New Jersey and Delaware Valley areas. Ironically, even though aviation has been an important part of Atlantic City's history, as low-cost air travel became accessible, Atlantic City's prominence as a vacation spot declined. As a result, it has been a challenge to attract airlines to Atlantic City; today, Spirit Airlines is the only scheduled airline serving the airport.

The people of Atlantic City have embraced aviation from its earliest days. The city and its citizens have also provided support, in times of peace and war, to the aviators who have flown the skies over their Boardwalk. Their generosity and patriotism were made clear in the dark days of World War II when they purchased the *Spirit of Atlantic City*, a P-47D Thunderbolt that was flown by one of the nation's greatest heroes and top aces, Walker "Bud" Mahurin. Other notable aviators, such as Medal of Honor recipient Thomas Hudner, record-setting Navy pilot Bob Dose, NJANG's David Haar, and countless others—some of whom have paid the ultimate price for their country—have contributed to the rich aviation history of Atlantic City.

One
AC Fledglings and First "Air Port"

ATLANTIC CITY AERO MEET, JULY 4–12, 1910. The world's first international flying meet took place in 1909 in France. Only a year later, the Aero Club of Atlantic City organized its aero meet. Noted aviators and officials of the meet are, from left to right, H.W. Sutton, manager; Henry M. Neely, contest chairman; Mrs. W. Ernest Shackelford; Arthur T. Atherhold; Glenn Curtiss; Clifford Harmon; Charles K. Hamilton; Lt. Hugh M. Willoughby; P.H. Deming; and aeronaut Augustus Post. (ACFLP.)

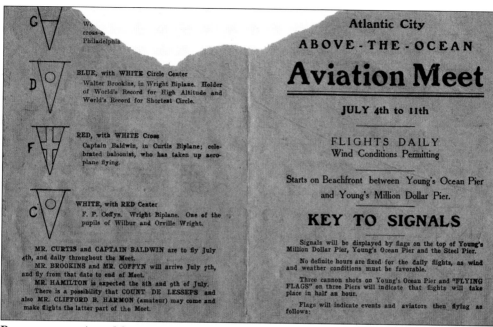

PROGRAM FROM AERO MEET. In 1910, flight of heavier-than-air machines was less than seven years old, and most Americans had never witnessed man fly or seen an aircraft. As described in the program shown here, the aero meet organizers relied on signal flags and cannon shots to inform the more than 100,000 viewers of the details of upcoming performances. (ACFPL.)

BROOKINS IN A WRIGHT AIRSHIP. Walter Brookins had the distinction in 1909 of being the first aviator trained by the Wright brothers for their exhibition team. Piloting a Wright Flyer at the Atlantic City Aero Meet, he became the first aviator to fly at an altitude of more than one mile on July 10, 1910. He also set records for transcontinental flight and endurance. (ACFPL.)

NOTED AVIATORS OVER THE BOARDWALK. The aero meet attracted virtually all the noted aviators of the day, two of which are flying over the Traymore Hotel. Starting as a boardinghouse in 1879, the Traymore became one of Atlantic City's premier resorts. Its popularity declined with that of Atlantic City in the 1960s, and it was razed in 1972. (ACFPL.)

"ABOVE-THE-OCEAN" AERO MEET. While Walter Brookins flies over the Boardwalk, fierce rival Glen Curtiss's aircraft is maneuvered on the beach. An aerodrome, which became known as Bader Field, was built off of the Absecon Channel to accommodate the participants; takeoffs were made from rails running along the beach, shown in this photograph, with landings occurring on the sand. (ACFPL.)

RECORDS BROKEN AT THE AERO MEET. On July 5, 1910, Glenn Curtiss made an 8.5-minute flight, half-a-mile out over the ocean—the first ever above the Atlantic. Four days later, Walter Brookins broke the altitude record, reaching 6,175 feet, but ran out of fuel before he could climb higher. He was able to safely glide the frail craft back to the beach in gusty winds. (ACFPL.)

Augustus Post, Noted Air-Traveler.

ATLANTIC CITY, THE NEW AIR PORT

By Augustus Post

IF you will add together the flying time of the two machines exhibited at the Atlantic City meet, July 4 to 12, 1910, you will see that the total time spent in the air by Messrs. Curtiss, Coffyn and Brookins was 6 hours, 39 minutes, 42¾ seconds. To get this, those in charge of the meet spent $25,000. This makes the cost of mechanical flight on this occasion something over $4,000 an hour, and that is about what it costs now, taking one event with another. More than grand opera prices, and the aviator can call himself one of the highest paid artists—for an artist he is, and, like Caruso, he is being paid now for a great deal that has taken place before he came on the stage, and, again like the artist, he is marketing the indefinable quality of genius.

How few of the thousands blackening the sands at Atlantic City realized the months and years of preparation for what seemed—as all supreme achievements seem—so easy, so entirely without effort! But try it; try even to think of doing it, and you will see how hard it can be; how much one must know, as well as how well one must be able to do. It is absolutely necessary to know the motor, all its little personal idiosyncrasies, its kinks and weak points as well as its strong ones. The aviator must know every bone in the body of his machine, just as a good horseman knows his horse, and have absolute confidence in every part, or if he cannot have that he must know its weak point, and how to favor that one point so that it will get the least strain possible. From this one begins to see what is required of the flying man, and can infer some of the instant and imperative demands on brain and body. Yes, the art and science of flight has called into being a company of men in whom certain traits of mind and person have been so wonderfully developed that at present they almost seem a race apart, "supermen" in one sense of the word at least. If man has at last made a new machine, the machine is certainly making a new man.

It was not mere curiosity that drew that immense assembly to the Jersey shore; although in such a meet as this there is the ever-present possibility of sudden and violent death; it was not the instinct for gladiatorial shows that riveted the attention of the multitude for so

THE FIRST "AIR PORT." At the time of the Atlantic City Aero Meet, flying was such a novelty that the term "aeronaut" was coined to describe the few fortunate "air travelers." Aeronaut Augustus Post avidly promoted early aviation through his writings and speeches. In an article for *Columbia* magazine, he coined the term "Air Port" to describe Atlantic City's prominence as a center of aviation. (ACFPL.)

CURTISS R-6 AND MF FLYING BOATS, 1919. Curtiss was one of the founders of the aviation industry and was licensed pilot No. 1. Glenn Curtiss returned to Atlantic City to establish a base off Brigantine Inlet for his flying boats. He also started an airborne sightseeing service, which catered to the wealthy residents of Atlantic City's elite hotels, using flying boats like the R-6 (left) and MF, shown here. (GEH.)

GLENN CURTISS WITH CURTISS MF. In the early years, Curtiss focused on naval aviation and especially on flying boats, such as this MF training plane built from 1918 for the US Navy. In civilian service, the MF was known as the Seagull. Powered by a 160-horsepower Curtiss K-6 engine, it could carry four people. (ACFPL.)

CURTISS H-16 FLYING BOAT, 1919. In 1914, Curtiss developed a then-giant flying boat for millionaire Rodman Wanamaker's transatlantic attempt. The outbreak of World War I cancelled this attempt, but Curtiss improved the design and sold it to the US Navy as the H-16, as shown here in Atlantic City, with its laminated wood veneer hull and two 200-horsepower Curtiss V-X-X engines. (ACFPL.)

CURTISS NC-4 ARRIVING IN ATLANTIC CITY, 1919. The U-boat menace of World War I stimulated the Navy's interest in antisubmarine aircraft. Curtiss responded with the NC family of flying boats. In May 1919, NC-4 became the first airplane to cross the Atlantic. The yacht *Delphine* tows NC-4 to the Atlantic City Yacht Club during the record-setting aircraft's recruiting tour after its return from Europe. (NNAM.)

CURTISS NC-4 AT ATLANTIC CITY. The May 1919 transatlantic attempt involved three Navy NC flying boats taking off from Naval Air Station Rockaway, New York, bound for Lisbon, Portugal. Only NC-4 successfully completed the flight. Making multiple stops next to Navy ships as well as points in New England, Canada and the Azores, the flying time was 26 hours and 45 minutes spread over almost 11 days. (NNAM.)

MAYOR BACHARACH GREETS NC-4. Sailors tie up NC-4 to the pier at Atlantic City Yacht Club as Atlantic City mayor Harry Bacharach greets Lt. Comdr. Albert Read and his crew. The epic flight had stops in Massachusetts, Nova Scotia, Newfoundland, then alongside 22 Navy warships to bridge the gap to a landing in the Azores, and a further 13 ships for the trip to Lisbon. (TH.)

ALBERT CUSHING READ, FLIGHT COMMANDER. Naval Academy class of 1907 graduate Albert Cushing Read commanded NC-4's crew of five naval aviators. NC-1, NC-3, and NC-4 left Rockaway, New York, but the first two aircraft could not complete the long Newfoundland-to-Azores leg. NC-2 was cannibalized for parts and never left New York. NC-4 now sits proudly in the National Naval Aviation Museum, Pensacola, Florida. (NNAM.)

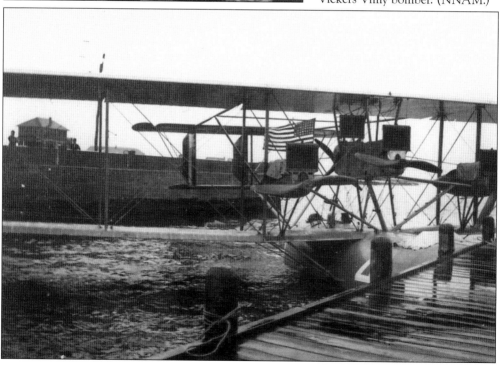

NC-4 AFTER THE PARTY. The record-breaking NC-4 sits forlornly on a rain-soaked Atlantic City Yacht Club pier while its crew is being "wined and dined" by city dignitaries. The Navy's accomplishment was overshadowed by the nonstop Newfoundland-to-Ireland flight by Britain's John Alcock and Arthur Brown just a few weeks later in a Vickers Vimy bomber. (NNAM.)

THE SPIRIT OF ST. LOUIS, OCTOBER 19, 1927. Even before Lindbergh's famous nonstop flight to Paris, New York millionaire Harry Guggenheim offered to sponsor a three-month publicity tour upon Lindbergh's successful return to stimulate interest in aviation. Flying the *Spirit of St. Louis*, Lindbergh visited 92 cities in all 48 states. Three stops from the end of the tour was Atlantic City's Municipal Airport, Bader Field. (ACFPL.)

GATHERING OF EAGLES, C. 1931. This historic photograph shows some of the most important aviation figures of the time visiting Atlantic City's Bader Field. Among them, standing in front of an Eastern Air Transport Curtiss Condor II, are Mayor and Mrs. Harry Bacharach, Amelia Earhart, Charles Lindbergh, and hotelier Adrian Phillips. Eastern Air Transport was formed from Pitcairn Airlines in early 1930, focusing on Eastern Seaboard travel. (ACFPL.)

FORD 5-AT-C TRIMOTOR, BADER FIELD, c. 1932. This Ford Trimotor, the 70th built, served with New York Airways, providing airline service linking New York (Newark Airport), Atlantic City, and Washington, DC. This particular aircraft flew with a number of US airlines in the 1930s and finally ended its flying days in Costa Rica. (ACFPL.)

HINDENBURG'S FATEFUL OVERFLIGHT. On May 6, 1937, the Nazi airship *Hindenburg* was headed to Lakehurst, New Jersey, from Frankfurt, Germany, but due to thunderstorms over Lakehurst, Capt. Max Pruss took the airship down the New Jersey Shore past Atlantic City until the weather cleared. Shown here over Atlantic City's Breakers and St. Charles hotels, the airship is heading back north to meet its fiery demise. (ACFPL.)

Two

Pine Barrens to Fighter Base

New Airport Announcement in Press. By 1940, the city administration realized the nation's first "air port," Bader Field, was in need of replacement. Start of construction of a new municipal airport about 10 miles west in Pomona, funded by the Work Projects Administration (WPA), was announced in November 1941—a month before Pearl Harbor was attacked. (Stan Ciurczak.)

AIRPORT TAKING SHAPE. Thousands of workers cleared 2,000 acres, laying out four one-mile-long runways. Looking northeast, the semicircular layout in the center would be the site of the terminal area. Still existing, Tilton Road crosses the photograph left to right while English Creek Avenue, crossing diagonally, was cut at the intersection when the Navy expanded the field in later years. (NA.)

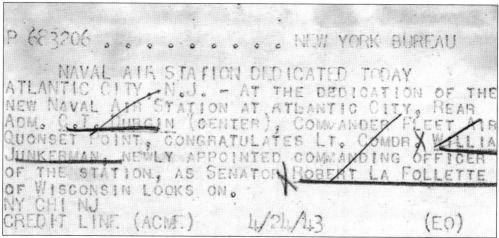

THE NAVY TAKING OVER THE PROJECT. Even before the outbreak of the war, the Navy was looking for air training sites on the Eastern Seaboard and took over the Pomona airport project in 1942. On April 24, 1943, construction was sufficiently complete to begin operations with the dedication of Naval Air Station Atlantic City, as noted in this telegram. (Stan Ciurczak.)

NAVAL AIR STATION COMMISSIONING. On April 24, 1943, Naval Air Station Atlantic City (NASAC) was commissioned into active duty according to message No. 061845 from the vice chief of Naval Operations to Lt. Comdr. W.J. Junkerman, who took command of the base on that day. Rear Adm. C.T. Durgin, Commander Fleet Air, Quonset Point, delivered the main address carried on Columbia Broadcasting's radio show *Spirit of '43*. SBD, TBF, and F6F aircraft of Carrier Air Group 25 (CVG-25) performed flybys for the occasion. This air group, established in 1943 at Norfolk, deployed on light fleet carrier *Cowpens* with TBF Avengers and F6F Hellcats after training at NASAC. In the photograph above, the woman with the fur shawl is Gertrude Gray, the mother of the Greate Egg Harbour Historical Society historian June Sheridan. (Both, GEH.)

SNJ of CASU-23, c. 1944. Carrier Aircraft Service Units (CASU) were created primarily to maintain carrier type of aircraft at naval air stations. By this organization, individual squadrons assigned to that base did not need to have their own maintenance cadres. CASU-23 was assigned to NASAC for the duration of World War II. This North American SNJ-4 was assigned to that unit. (NNAM.)

Flight of CASU-23 SNJs Over New Jersey. Originally built as the private venture NA-16 basic trainer, contract production continued from 1935 through 1944. Known as the T-6 in the Army Air Corps, the SNJ in the Navy, and Harvard in Royal Air Force service, more than 16,000 were built by North American Aviation in the United States. Hundreds more were built under license in other countries. (NNAM.)

LT. STANLEY VEJTASA SELECTED TO SET UP FIGHTER SCHOOL. Battle of Coral Sea ace Lt. S.W. "Swede" Vejtasa was chosen to set up the Fighter Training Unit's curriculum. Shown in this July 1942 photograph, Comdr. James Flatley (left) congratulates Vejtasa for shooting down seven Japanese aircraft in one mission; Lt. John Leppa looks on. Along with eight other combat veterans, his team converted NASAC to a fighter-only base. After the war, Vejtasa continued flying and rose in rank, commanding *Constellation* and the Navy Pacific Fighter Force during Vietnam. Officially credited with 11 victories, he earned many decorations, including three Navy Crosses. (Above, NA; right, NNAM.)

NAVY V. MOTHER NATURE, ROUND ONE. One attraction for selecting the Pomona site for the new municipal airport, and later the naval air station, was the favorable weather compared to other sites. However, it was certainly not immune to bad weather, as shown in this series of photographs of a sand storm that engulfed the base. Looking southeast, the two hangars and flight line are almost swallowed by blowing sand. (NA.)

FLIGHT LINE IN SAND STORM, C. 1944. Here, the ramp is shown from the top of one of the two hangars. In the center are silver SNJs assigned to CASU-23. On the left, F6F Hellcats are lined up with their wings folded back next to their fuselage—a trademark of Grumman. Far right is a trio of F4F Wildcats. In the background are Vought F4U Corsairs. (NA.)

GRUMMAN HELLCATS WEATHERING THE STORM. During the war, Grumman F6F-3 and F6F-5 Hellcats were assigned to numerous fighter squadrons at Atlantic City. The "-5" offered a few refinements based on combat experience. Starting in 1943, aircraft were repainted with non-specular dark blue for top surfaces, a lighter gray-blue for lower sides, and white undersides, as shown here. (NA.)

NAVAL AIR STATION RAMP AND TOWER. Navy linemen have secured these SNJs and F6Fs against the sand storm, including tie downs and covers for the delicate pitot tubes on the starboard wingtips. The air station control tower is barely visible in the background. (NA.)

NAVY V. MOTHER NATURE, ROUND TWO. On September 13 and 14, 1944, Atlantic City suffered a direct hit from a hurricane. Although the air station itself avoided substantial damage, the surrounding area was not so fortunate. The Brigantine Hotel, taken over by NASAC in 1943 for the training of Combat Information Center radar controllers, was heavily damaged, losing all windows on the ocean side. (GEH.)

AFTER THE STORM. This panorama of the air station buildings and ramp was taken after the September 1944 hurricane. The air station was relatively unaffected, but about 500 trees were uprooted, and many of the buildings were flooded. The ramp is dominated by rows of F6F Hellcats. (GEH.)

NASAC, August 1944. This aerial photograph shows how the air station had developed in the 1.5 years since its activation. The top of the photograph is roughly in the north direction. The semicircular taxiway around the proposed site of the municipal airport terminal is clearly seen. (NA.)

F6F Hellcats of VF-86, 1944. Fighting Squadron 86 (VF-86) trained at Atlantic City during 1944 before assignment to fleet carrier *Wasp* (CV-18) in time for the Battle of the Philippine Sea. Grumman built 12,275 Hellcats in Bethpage, Long Island, located about 120 miles northeast of NASAC. (NNAM.)

HELLCAT FROM VF-74 LAUNCHING DURING OPERATION ANVIL. Fighting Squadron 74 (VF-74) trained at NASAC before joining escort carrier *Kasaan Bay* for the August 1944 invasion of South France, code-named Anvil. Along with VOF-1, VF-74 was tasked with close air support and interdiction missions. (NNAM.)

HELLCAT COMING TO GRIEF ON TULAGI. Navy's Observation Fighter Squadron 1 (VOF-1) was activated at NASAC in December 1943. VOF squadrons were tasked with directing naval gunfire and VOF-1 on escort carrier *Tulagi* fought alongside VF-74 in Operation Anvil. Although 11 Hellcats from the two squadrons were lost mainly to ground-fire, eight Luftwaffe aircraft were claimed. (NNAM.)

F4U-1Ds of VF-89 Over New Jersey. Fighting Squadron 89 was established at NASAC in October 1944. Early in 1944 the Navy adopted an overall midnight-blue color scheme, shown here, for carrier aircraft, replacing the three-color scheme, as shown on page 27. (Vought.)

F4U-4 of VF-82. Fighting Squadron 82 (VF-82), formed at NASAC in April 1944, was one of the first squadrons to receive the updated F4U-4 version of the Vought Corsair. Aircraft assigned to carriers were organized into Carrier Air Groups (CAGs), consisting of fighter, dive/scout bomber, and torpedo bomber squadrons. The individual types were assigned to different air stations with NASAC being the home for fighters. (Vought.)

MARTIN JM-1 MARAUDER OF VJ-4, 1944. Fighter pilot training included air-to-air gunnery practice. The Navy acquired 272 Martin Marauders assigned to utility squadrons (coded VJ) to serve primarily as target tugs. During various periods of the war, NASAC was home for VJ-1, VJ-4, and VJ-7. The JM-1s were ex-USAAF AT-26B trainer conversions of the operational Marauder B-26 medium bombers. The high accident rate in the early years of Army Air Corps service earned it the name "Widow Maker." Due to the relatively short wingspan, as seen in these photographs, it also was called the "Flying Prostitute" since "it had no visible means of support." In operational use, it actually had a lower accident rate than comparable aircraft. (Both, NA.)

JM-1 Target Tug of VJ-4. For training fighter pilots in air-to-air gunnery, the Martin Marauder target tug was connected by a long cable from a winch in the tail of the aircraft to a specially built target, which could be a "banner" or a glider. The connection point is shown in the photograph at right. In the case of a glider type of target, the tow aircraft would takeoff pulling the target, as shown in the photograph below. With the target towed a suitably safe distance behind the Marauder, fighter aircraft would initiate firing passes from a direction to minimize any danger of shooting down the JM-1. To minimize this danger, the Navy painted its target towing aircraft a bright yellow-orange color. (Both, NA.)

NORTH AMERICAN SNJ-5 TEXAN. In addition to the target towing aircraft, the air station also had a number of other utility aircraft assigned for general use. This pristine SNJ-5 Texan trainer was one of more than 1,300 produced for the Navy. It was comparable to the Army Air Corps's AT-6D trainer. (NNAM.)

CROWDED NASAC RAMP, AUGUST 1945. This photograph, taken at the end of World War II, shows aircraft of Fleet Carrier Air Groups 4, 45, and 97. F6F-3 and -5 Hellcats, with their wings folded back along their fuselages, predominate. To the right is a row of F4U-4 Corsairs, and in the background are silver SNJs and TBF/TBM Avengers. (TH.)

NASAC, August 1945. These aerial photographs show how NASAC looked at the end of the war. Comparing these photographs with earlier ones shows the degree to which the pine forest was converted to an air station. The above photograph, looking north, shows the ramp, hangars, and many barracks, shops, and classroom buildings toward the lower right, with the Great Bay in the distance. In the below aerial, with west toward the top, the semicircular area is the location chosen for the municipal airport before the Navy took over the project. During the war, medical casualty flights brought countless wounded soldiers to Atlantic City for treatment at the Army's England General Hospital (converted luxury hotels); some of the C-47s used for this purpose are shown on the semicircular ramp. The air station hangars and ramp are to the left. (Both, NA.)

GARDINER'S BASIN. With many training flights from NASAC over the ocean, a means of rescuing downed airmen from the sea was needed. A number of crash rescue boats for this purpose were stationed at a Gardiner's Basin marina; one is shown in the center of the photograph. Nestled on a protected channel across from the state marina and Coast Guard station, "The Basin" was once a haven for rumrunners and commercial fishermen. (NA.)

OLF CLARK FIELD, AUGUST 1945. An Outlying Landing Field that was controlled by NASAC, known as Clark Field, was scratched out in Ocean City, New Jersey. With two short runways, the field was used as an emergency diversion and landing practice field. Located 13 miles south of NASAC, the field is today the Ocean City Municipal Airport. (NA.)

BADER FIELD, AUGUST 1945. Starting in April 1944, Bader Field in Atlantic City was used as an OLF. To relieve the congestion from so many aircraft based at the main air station, OLFs were relied upon to provide additional runways for practice takeoffs and landings. The Atlantic City skyline, Boardwalk, and famous Steel Pier can be seen at the top of the photograph. (NA.)

OLF WOODBINE, AUGUST 1945. Woodbine Airport, located 20 miles southwest of the NASAC was perhaps the most important of its Outlying Landing Fields (OLF). Originally an OLF of Naval Air Station Wildwood, it was transferred to NASAC in 1944. With three 2,500-foot runways, it was used for Field Carrier Practice Landings (FCPL), where pilots honed their skills at landing on an aircraft carrier. (NA.)

BARRAGE BALLOONS. Although not directly related to NASAC, nearby Seyfang Laboratories were an important part of Atlantic City's wartime effort, producing antiaircraft barrage and weather balloons. Wartime employment of women is clearly evident in these photographs. Although mostly associated with the protection of Britain from Luftwaffe attack during the Blitz, barrage balloons made by Seyfang and others were used to protect American cities (particularly, West Coast oil fields, coastal refineries, and armament factories) and convoys. In addition, the D-Day invasion fleet also benefitted from barrage balloon protection against air attack. The helium-filled balloons were tethered by steel cables, sometimes 5,000 feet long. While the balloons would be visible, the cables were not, forcing the attackers to stay well above them. Seyfang weather balloons were also widely used, and some believe one was the "UFO" of the infamous Roswell incident in 1947. (Both, ACFPL.)

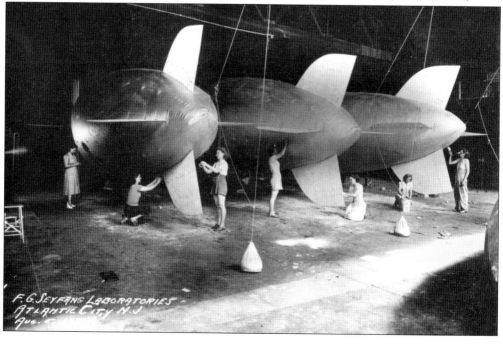

Three
POSTWAR TECHNICAL DEVELOPMENTS AND KOREA

NASAC Ramp, 1948. Postwar, the mission of NASAC was changed from training fighter pilots to providing support for fleet activities by adding attack, antisubmarine, reconnaissance, and fighter squadrons. The ramp is crowded with TBMs from Attack Squadron 1 Lima (VA-1L) and F4Us of Fighting Squadron 1 Lima (VF-1L), squadrons from Light Fleet Carrier Air Group 1 noted by "SA" tail codes. A Convair PB4Y Privateer is in the background. (TH.)

"Sub Hunting" Over the Boardwalk. Two VA-1L TBM-3E Avengers are shown flying south over Atlantic City and the Boardwalk. The submarine-searching radar pods under their starboard wings are clearly seen, as is the Steel Pier jutting out from the beach. The photograph was taken from the observer's "belly" window of a third TBM. (NNAM.)

TBM-3E Avenger, c. 1948. VA-1L was originally established as Torpedo Squadron 58 (VT-58) in 1946, but the Navy deleted the torpedo bomber function and replaced it with the attack designation. The photograph shows TBM-3E Avenger antisubmarine aircraft of VA-1L over Naval Air Station Atlantic City. This version carried a search radar pod under the starboard wing and became the chief postwar frontline version of the venerable Avenger. (TH.)

Drones Prepared for Atomic Bomb Tests. These colorful aircraft are some of the F6F-3K and -5K Hellcats modified for use as pilotless drones, as denoted by the "K" suffix. The aircraft were tested at NASAC prior to their 1946 use in Operation Crossroads—the atomic bomb tests at Bikini Atoll. Sophisticated radio-control equipment was installed to allow pilotless takeoffs, flight, and landings. Since missions could involve multiple aircraft, they were color coded according to the controlling radio frequency. In the right photograph, a lineup of many of the 18 aircraft assigned to NASAC is shown, with the control tower in the background. In the photograph below, the Hellcat in the foreground has a particle collection box fitted inside of its port landing gear. Some of them also have an extended tail wheel strut, a modification to make radio-controlled taxiing and takeoffs easier. (Right, TH; below, NA.)

RADIO-CONTROLLED HELLCATS. The F6F-3K and -5K drones could be controlled from a ground station or from aircraft flying alongside. During the development and testing of the drones, it was common to control the takeoff and landings from a ground station (above) but control the flight portion from a nearby aircraft (below) since the range from the ground control point was limited. The drone technology tested at NASAC was also used during the Korean War where radio-controlled, bomb-laden Hellcats were flown off carriers to attack heavily defended targets. One might say that the unmanned aerial vehicles (UAV), commonly called "drones" in the news today, are a development of the work at NASAC. (Both, NA.)

Atlantic City Drones Deploy to Pacific Tests. The Hellcat drones (shown on the flight deck) were loaded on *Shangri-La*, which transited the Panama Canal to join the Operation Crossroads test fleet at the Marshall Islands' Bikini Atoll. They were catapulted from the carrier and flown through the mushroom cloud. Wisely, recovery was on a Roi Island airstrip rather than a hazardous radio-controlled carrier landing of a radioactive Hellcat. (NA.)

Testing of F6F-3Ks for Radioactivity. The boxes mounted under the wings of the F6F-3K Hellcats contained filters to trap radioactive particles as the aircraft flew through the radioactive cloud. After landing on Roi Island, the exposed aircraft were then treated with great caution and monitored for radioactivity, as shown here. (NA.)

REMOVAL OF SAMPLE BOX. The photograph shows the careful removal of the sample collecting filter box for analysis, showing obvious caution in handling the potentially dangerous sample box. The Hellcat now displayed at the National Air and Space Museum is one of the former NASAC drones. (NA.)

AM-1Q OF VC-4. Composite Squadron 4 (VC-4) was the night/all-weather fighting squadron established at NASAC. The Martin AM-1 Mauler was developed to the same Navy requirement as the Douglas AD Skyraider but was less successful. Most of those built ended up in the reserves. VC-4 received eight of the electronic warfare version, the AM-1Q, one of which is shown flying off *Kearsarge* during 1949 carrier qualifications. (NA.)

VC-4 Corsair on Antietam. Composite Squadron 4 was a large squadron and was assigned tail code "NA." The squadron typically deployed four aircraft detachments to aircraft carriers to provide night-fighting capability—a skill not highly developed at the time. This VC-4 F4U-5NL (the "N" denoting "night" and "L" denoting "winterized") is nearing the starboard catapult of Antietam on one of its deployments. (Vought.)

VC-4 and VC-62 Vought Corsairs, 1949. The Navy relied on fleet composite squadrons to provide detachments of aircraft to carriers at sea for specialized expertise and mission capabilities. This photograph shows a mixed formation of photographic reconnaissance F4U-5Ps of VC-62 (tail code "TL") and night fighter F4U-5Ns of VC-4. (NNAM.)

VC-4 Detachment on Leyte. On June 25, 1950, North Korea invaded South Korea starting a three-year "police action" in which naval aviation would play a pivotal role. Detachments from NASAC's VC-4 and VC-33 made multiple deployments. VC-4 sent a four F4U-5N/NL Corsair detachment on Leyte's first Korean War cruise from September 6, 1950, to February 3, 1951, shown here; Corsairs of VF-32 are readied on the catapults. (NNAM.)

Corsair Recovering on Bon Homme Richard. VC-4 and VC-33 sent detachments on Bon Homme Richard's second Korean War cruise from May 20, 1952, to January 8, 1953. While VC-4 was a night-fighting unit, in Korea, it was used exclusively for more dangerous night attack, similar to the task of VC-33. For landing operations, all aircraft are parked as far forward on the carrier's deck as possible, as shown in this photograph. (NNAM.)

CORSAIR OF VC-4 FLIES OFF BON HOMME RICHARD. These photographs show a VC-4 F4U-5NL Corsair unfolding its wings and flying off the carrier's deck during its Korean War detachment. While the carriers of that period were equipped with catapults, when the deck park allowed it, it was common for the propeller aircraft to fly off the deck directly. The pod containing the APS-5 radar on the starboard wing is clearly seen, as are the baffles blocking the exhaust flame from interfering with the pilot's night vision. The "L" suffix on the designation indicates "winterized," in the form of wing leading edge deicer boots—very necessary during Korea's notoriously cold winters. Although a night-fighting unit, when required by air group demands, VC-4 also participated in daylight operations. (Both, Vought.)

Skyknight of VC-4, 1952. NASAC's VC-4 was a large squadron equipped with different types of specialized aircraft. Besides the Corsairs, the unit was equipped with Douglas F3D-2 Skyknight and McDonnell F2H-2N Banshee night fighters. These photographs show one of the unit's Skyknights during carrier qualifications on *Franklin D. Roosevelt*. VC-4 was the only Navy squadron to deploy the Skyknight on a cruise, sending detachments to both Atlantic and Pacific fleet carriers. Their assignment was to afford all-weather and nighttime protection of the fleet from air attack. Production of the F3D-2 totaled 237, many of which went to the Marines who continued to use them during the Vietnam War. (Both, NA.)

Exercise Mainbrace, 1952. VC-4 F3D-2 Skyknights and F2H-2B "special mission" (i.e., nuclear bomber) Banshees along with VC-33 AD Skyraiders participated in Exercise Mainbrace, NATO's first large naval maneuvers. More than 80,000 men, 1,000 aircraft, and 200 ships participated in the simulated defense of Europe's Northern Flank from Soviet Pact attack. These photographs show VC-4 Skyknight and Banshee aircraft (denoted by "NA" tail code) among other squadron aircraft at NAS Norfolk before embarkation on carriers *Midway* and *Franklin D. Roosevelt* for the September 14–25, 1952, exercise. A total of 10 aircraft carriers participated, including three from the United Kingdom and one from Canada. (Both, NA.)

VC-4 SKYKNIGHT GOING TO WAR. VC-4 sent F3D-2 Skyknights and F2H-2B Banshees (again joining Skyraiders of sister-squadron VC-33) on *Lake Champlain* for a Korean War cruise from April 26 to December 4, 1953. Shown above, a Skyknight is prepared at NASAC before deployment, while the photograph below shows the forward deck with a Skyknight on the right, next to a VC-4 Banshee ("NA" tail code). As Skyknights were large aircraft (often called the "Whale") with a disruptive nighttime mission, the ship's captain sent them to airfield K-6 in Korea, joining a similarly equipped Marine Squadron (VMF[N]-513). The F2H-2B was a specialized Banshee with stiffened wings and hardpoints giving the capability of delivering "special weapons," the euphemism for nuclear bombs. No doubt the carrier had such weapons aboard, as confirmed by reports of Marines guarding the ship's magazine around the clock. (Both, NA.)

CORSAIRS OVER JERSEY SHORE, SEPTEMBER 1954. The "bent-wing" Corsair is perhaps the most recognizable icon of naval aviation in the 1940s and 1950s. It was in production longer than any other fighter of the World War II era. It had an 11-to-1 kill ratio against Japanese aircraft and was the last piston engine fighter produced for the US armed forces. For NASAC, the Corsair played a prominent role from its establishment in 1943 until the Corsair was finally retired from frontline squadrons in 1954. The photograph above shows a division of VC-4 F4U-5N Corsairs heading south over the Ocean Drive causeway crossing the Great Channel to Stone Harbor, New Jersey. In the photograph below, the division makes a banking turn to the north, with the Atlantic City skyline visible toward the top right. (Both, NA.)

CORSAIRS BID FAREWELL TO NASAC. Vought Corsairs flew from NASAC during World War II, assigned to many of the fighter squadrons trained for deployment overseas. After the war, Corsairs flew with Composite Squadron 4 (VC-4) as well as Air Development Squadron 3 (VX-3). Most of the major models, starting from the F4U-1, through the F4U-4, and finally the F4U-5N, shown here, flew from NASAC's runways. It is only fitting then that the Corsair was retired from active duty in the Navy by VC-4 in 1954. The photograph above shows a line-abreast formation flying north, parallel to the shoreline opposite Atlantic City and Bader Field. The left photograph is symbolic of its retirement from NASAC and the Navy, as the aircraft "peel off" from the camera offshore Brigantine, New Jersey. (Both, NA.)

NASAC Becomes a Master Jet Base. Even before the retirement of the Corsair, NASAC's squadrons were being re-equipped with jet aircraft. By 1954, when these photographs were taken, the McDonnell F2H-3 and -4 Banshees had become the Navy's premier all-weather/night fighters. VC-4 was equipped with both versions, with the -3 and -4 models differing in the radar (Hughes APQ-41 v. Westinghouse APG-37) and upgraded engines. The photograph above shows a division of F2H-4 Banshees flying north offshore of the Atlantic City Boardwalk opposite Bader Field. In the photograph below, a gaggle of Banshees flies southbound offshore Ocean City, New Jersey. The airfield toward the top center was Clark Field, an Outlaying Landing Field used by NASAC during the war. With runways too short for Navy jet aircraft, the airfield became Ocean City Municipal Airport. (Both, NA.)

"BIG BANJOS" SIGHTSEEING OVER NEW JERSEY, 1954. As is common to all military branches, nicknames are applied to people as well as equipment. When the F2H-1 and -2 Banshee was introduced to the Navy, it gained the sobriquet "Banjo." The later F2H-3 and -4 Banshee had an eight-foot fuselage extension compared to the earlier models, as well as a revised tail. Naturally, the larger versions became known as "Big Banjos." In the photograph above, a division flies over Absecon Channel, with Brigantine to the right and Atlantic City to the left. The photograph below shows a F2H-4 heading south over the Boardwalk and the Steel Pier. NASAC is discernible just to the right of top center. (Both, NA.)

AD Skyraiders of VC-33, 1950. When the Navy decided to establish its eastern center to develop the capability for night and all-weather warfare at NASAC, two squadrons were assigned—Composite Squadrons 4 (VC-4) and 33 (VC-33). This lineup of AD-2Qs of VC-33 on the NASAC ramp sits in front its predecessor, a TBM-3E. VC-33 deployed AD-4N night attack detachments on four Korean War cruises. (NA.)

VC-33 Skyraider, 1950. VC-33 was originally established as an antisubmarine squadron at Norfolk Naval Air Station. But when it was transferred to Atlantic City in June 1950, it gained a new mission—night attack. The predominant squadron aircraft assigned were various versions of the AD Skyraider. Composite squadrons (like VC-33 with the "SS" tail code) deployed detachments to carriers. This AD-4 is shown beginning its takeoff run from *Coral Sea*. (NNAM.)

HAZARDOUS CARRIER OPERATIONS. Naval aviation accident rates in the early 1950s, flying high-performance aircraft from World War II vintage carrier decks, were horrendous. Here, a VC-33 AD-4 stalls into the sea as a result of a "cold cat"—the malfunction of a H4B hydraulic catapult on *Tarawa*. Fortunately, the Navy embraced many British carrier innovations, including steam catapults, angled decks, and a mirror landing system, which drastically reducing accident rates. (NNAM.)

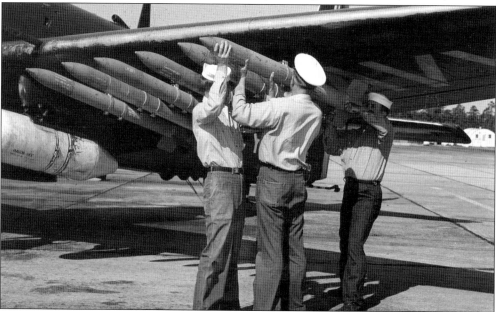

VC-33 GROUND CREW WITH ROCKETS, JULY 1951. Although justly proud of their own capabilities and accomplishments, Navy pilots are the first to give credit to their maintenance and ground crews who labor tirelessly to keep them in the air! Loading munitions onto an aircraft is a meticulous and laborious task. In this photograph, five-inch high-velocity rockets are manually loaded onto a VC-33 AD Skyraider. (NA.)

VC-33 SKYRAIDERS, C. 1952. The Douglas AD Skyraider family of aircraft is among the most successful and versatile aircraft ever developed. Originally designed as a carrier-based dive and torpedo bomber, wartime experience stressed the importance of the ability to carry large payloads over great distances. The result was a large, powerful aircraft with no provision for internal storage of weapons. It was built in seven main versions but was modified to perform other functions, such as night attack, antisubmarine warfare, early warning detection, and electronic countermeasures. The photograph above shows a VC-33 AD-4Q, one of 39 of this two-seat (note the door and porthole for the operator) electronic countermeasure version built. The photograph below is a night attack AD-4N, with large radar pod under its wing, as deployed for VC-33's three Korean War cruises. (Both, NNAM.)

VC-33 AD-5Q Skyraiders. In the Skyraider attack versions, ordnance totaling nearly 10,000 pounds could be hung from 15 under-fuselage and under-wing pylons, giving the aircraft the ability to deliver more bombs from a carrier deck than the World War II B-17 Flying Fortress bomber. It is even credited with MiG kills during the Vietnam War. However, later in its service career, its specialized roles, such as electronic countermeasures, became more important. The AD-5 resulted from a major redesign. The fuselage was widened to accommodate side-by-side seating and lengthened to accommodate equipment, and the tail was enlarged to improve stability. The modification for electronic countermeasures provided for a crew of four and the necessary equipment. In the photograph above, VC-33 AD-5Qs are prepared for a sortie, while in the photograph below, three Skyraiders fly down the shoreline south of Atlantic City. (Above, NNAM; below, NA.)

VC-33 PERSONNEL DEPLOYING FOR CRUISE, 1950. Deploying a Composite Squadron detachment involves more than just flying the aircraft from the home station to the aircraft carrier. It also involves moving ground personnel and equipment required to maintain the aircraft. This photograph shows VC-33 personnel loading onboard one of the Navy's two R6V Constitution aircraft for deployment on *Coral Sea*. (NA.)

AIR DEVELOPMENT SQUADRON THREE LINEUP, 1950. The third important postwar squadron established at NASAC was Air Development Squadron 3 (VX-3). Formed by the merger of squadrons VA-1L and VF-1L on November 20, 1948, its mission was to develop tactics for employment of the latest Navy aircraft. From the top, the aircraft are the McDonnell F2H-1 Banshee, McDonnell FH-1 Phantom, and the Chance Vought F6U Pirate. (NA.)

VX-3 CARRIER QUALIFICATIONS, C. 1950. Part of VX-3 responsibilities was to develop tactics and prove procedures "around the boat," Navy parlance for carrier operations. This photograph shows VX-3 Vought F4U-5 Corsairs unfolding their wings in preparation for flying off the bow of the carrier, with squadron F2H-2 Banshees, wings still folded, to the right and F9F Panthers behind them. Note the squadron's "XC" tail code. (TH.)

VX-3 HANGAR, NASAC, 1950s. Air Development Squadron 3 (VX-3) ramp shows blue F9F-8 swept wing Cougars, straight wing F9F-5 Panthers, and silver TV-2 Sea Stars in front of the hangar. To the right are AD Skyraiders with folded wings. Most of the buildings shown have since been razed; the ramp is now the home to F-16Cs of the New Jersey Air National Guard. (TH.)

VX-3 RAMP. This photograph of the VX-3 ramp displays the variety of aircraft operated by the squadron at the same time. The aircraft types are, clockwise from top: Beech SNB Navigator trainer; Douglas AD Skyraider; McDonnell F2H-2P Photo Banshee; F2H-3 "Big Banjo" Banshee; Chance Vought F7U Cutlass; Grumman F9F-8 Cougar; F2H-2N Night Banshee; Lockheed TV-2 Sea Star trainer; and in the center is a Douglas F3D Skyknight. (TH.)

VX-3 INVENTORY C. 1954. The diversity of aircraft flown by VX-3 from NASAC runways was amazing as many types were being introduced into service. The echelon from the top shows a swept wing Grumman F9F-8 Cougar and its straight wing Panther predecessor; a silver McDonnell F2H-3 Banshee alongside its predecessor, F2H-2 Banshee; and F2H-2P Photo Banshee, followed by a Douglas AD Skyraider. Most VX-3 pilots were qualified to fly all types. (TH.)

GRUMMAN F9F-6 COUGAR, 1954. Grumman was very successful at developing its first jet fighter, the Korean War veteran F9F-2 and -5 Panther, into an excellent swept wing derivative: the Cougar. Despite the same F9F designation, it was actually a different aircraft, hence a different name. More than 700 of the F9F-6 version were built. This VX-3 Cougar is flying over New Jersey. (TH.)

VX-3 BANSHEE CARRIER QUALIFICATIONS. This F2H-2 Banshee of VX-3 ("XC" tail code) is about to "trap" (i.e., land) on light fleet carrier *Saipan*. The tailhook is down, canopy open, and the pilot has just "taken the cut" from the landing signal officer (LSO), indicating his approach was satisfactory. (TH.)

PHOTO BANSHEE APPROACHING CARRIER. This VX-3 F2H-2P photorecon Banshee is approaching a carrier for landing. The carrier's wake is behind the Banshee. The "plane guard" destroyer, seen below the Banshee, typically follows closely behind the carrier ready to rescue any flyers that might end up in the sea. (TH.)

F2H-2 COMING TO GRIEF, MARCH 11, 1953. Lieutenant Commander Vescovie flying a F2H-2 Banshee of VX-3 suffered a collapsed landing gear and crashed landing at NASAC. Since pilots say, "Any landing you can walk away from is a good one," this is a good one! Damage seems to be minor and repairable. (TH.)

"BIG BANJO" OF AIR DEVELOPMENT SQUADRON 3. The McDonnell F2H-3 and -4 were the ultimate versions of the Banshee. Designed for all-weather and night interception of hostile aircraft, VX-3 developed tactics and procedures prior to the aircraft's release to fleet squadrons. The Navy tried the unpainted aluminum finish during the 1950s. (TH.)

VX-3 SKYRAIDER OVER NEW JERSEY. The Skyraider was one of a number of types of new Navy tactical aircraft that was evaluated by VX-3 in the late 1940s and early 1950s. One of the most versatile aircraft ever produced, it served the Navy, Marines, Air Force, Royal Navy, France, Sweden, and numerous smaller air forces for many years after production ended. (TH.)

Douglas AD-4 Skyraider Of VX-3 on a Photograph Shoot Over NASAC. The photograph was taken from a squadron Beech SNB Navigator trainer, popularly known as "the Bug Smasher." The fast closure rate between the AD and SNB is the reason for wing up and speed-brakes out configuration of the "Able Dog," the early nickname for the AD series of attack aircraft. (TH.)

F7U-3 Cutlass of VX-3, October 1954. The unconventional Chance Vought Cutlass benefitted from wartime German research into tailless aircraft. Pitch-and-roll control was from combined elevators/ailerons ("elevons") on the wing, while vertical stabilizers were located on the wings at the end of the center section. The first flight of the prototype XF7U-1 was in September 1948. (NA.)

F7U-3s of VX-3 Overfly Atlantic City. The performance of early F7U-1s was poor, requiring massive redesign resulting in the definitive production version, the F7U-3. VX-3 evaluated the F7U-3, F7U-3P photorecon, and F7U-3M missile-carrying fighter versions. Advanced aerodynamics was betrayed by insufficient thrust from the two Westinghouse J-34 engines (a common ailment of early 1950s jets), leading to the nickname "Gutless Cutlass." Pilots complained that the Westinghouse engines "put out less heat than the company's toasters." The unpainted aluminum Cutlasses, shown in these photographs, are the standard fighter version. In the photograph above, the extant Boardwalk Center is seen; the pier has been replaced with an over-water mall. The left photograph shows a Cutlass over the current site of the Trump Taj Mahal Casino, with the Absecon Inlet in the background. (Above, NA; left, TH.)

CRUSADERS OVER THE JERSEY SHORE, 1956. Perhaps the most exciting and successful 1950s naval fighter was the Chance Vought F8U Crusader. After the success of the wartime F4U Corsair, Vought stumbled with the F6U Pirate and F7U Cutlass. They regained their prominence, however, with the design of the Crusader. Built around a Pratt & Whitney J57 afterburning turbojet engine, it was fitted with a high-mounted variable incidence wing that allowed a reduced approach speed for carrier landings. In the photograph above, a division of F8U-1 Crusaders goes "feet wet" over Great Egg Harbor Inlet, with Margate and Ventor to the left and Ocean City to the right. The photograph below shows a Crusader flying north over Strathmere, New Jersey. (Both, NA.)

WELCOME TO JERSEY, 1956. The prototype Crusader first flew on March 25, 1955, breaking the sound barrier on this flight. The Navy now had a carrier fighter equal in performance to its Air Force counterparts. The Crusader in this photograph, retaining the markings from the Patuxent River Test Center where initial suitability testing took place, banks toward the lower end of Long Beach Island and Mystic Island. (NA.)

CRUSADERS "AROUND THE BOAT." VX-3's evaluation of new aircraft included continued testing and procedures development onboard aircraft carriers, including numerous periods at sea. Here, a number of VX-3 Crusaders prepare to launch from *Franklin D. Roosevelt* during carrier qualifications. Note the leading edge of the wing raised for takeoff and landings—a unique feature of the Crusader design. (Vought.)

VX-3 Panther, July 1951. From the first days of carrier aviation, landings were guided by the landing signal officer (LSO). Until the mid-1950s, the LSO used hand signals and radio communications to provide height and lineup information to the pilot. Here, a F9F Panther receives the "cut" signal from the LSO at NASAC during Field Carrier Practice Landings (FCPL), a term still in use today. (NA.)

Mirror Landing System. While pilots making carrier landings still rely on the assistance of the landing signal officer, guidance in terms of alignment and height are now provided by a gyroscopically controlled mirror landing system. A British innovation, VX-3 tested and perfected an early version. This photograph shows a squadron F3H Demon trapping on *Saratoga* using the mirror and colored datum lights shown in the foreground. (NNAM.)

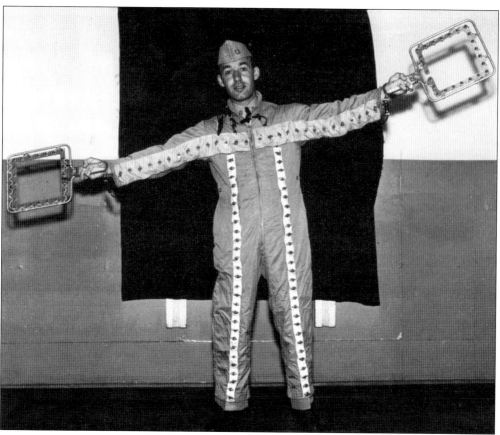

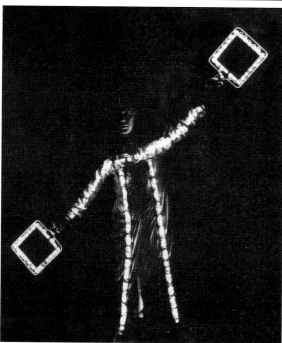

Night LSO Suit, 1952. Prior to the innovation of the mirror landing system, the LSO gave visual signals to approaching pilots using "paddles" as shown in these photographs. This system, originated in the 1920s, was adequate for clear, daylight conditions. However, as the Navy developed night and all-weather capability, something better was needed. The first attempt was ultraviolet light illumination of the LSO, but this proved impractical. In 1952, NASAC based VC-4's Lts. K.C. Pailer and T.S. Lockard applied strings of Christmas tree lights to their suits and paddles, giving pilots both vertical and horizontal reference. In the photograph above, Lieutenant Lockard demonstrates signaling in daytime, while in the left photograph, the same signal is given at night using the "Christmas tree" suit. (Both, NA.)

AIR-TO-AIR REFUELING TRIALS, C. 1954. To extend the range of tactical aircraft and provide support for returning aircraft short of fuel, air-to-air refueling has become a standard procedure. Part of VX-3's responsibility was developing such procedures for aircraft entering the fleet. North American designed the AJ Savage as a nuclear bomber at a time when the dimensions and weight of nuclear weapons required such a large aircraft. It featured two piston engines augmented by a turbojet in the tail. Development of lighter, compact nukes made the Savage's attack role superfluous, and many were converted to aerial refueling tankers. Above, a Grumman F9F-7 takes on fuel from a North American AJ-2 Savage of VC-8. The F9F-7 was the penultimate variant of the Cougar, similar to the F9F-6 but with an Allison J-33 engine. Below, a Cutlass refuels from a VC-7 Savage. (Both, TH.)

NASAC TOWER AND BASE OPERATIONS, C. 1954. The tower and base operations building, shown here, was the control center for air station activity. It is festooned with a "forest" of aerials and antennas, indicative of the growing importance of electronics and communications. This structure is gone as are most buildings from the Navy days. (TH.)

NASAC, OCTOBER 1947. This aerial photograph, taken from 8,000 feet, shows the air station basically as it looked after the end of World War II. With the top of the photograph aimed toward the east, the four runways, curved apron of proposed municipal airport site, and many structures of the air station, are clearly evident. Few aircraft were on the ramp compared to the hundreds during the war. (NA.)

NASAC BECOMING A MASTER JET BASE. These aerials, taken in 1952 (above) and 1954, show the changes that were made to accommodate the needs of the Navy's new high-performance jet aircraft. In 1952, work at extending the runways from their original 5,000-foot length was underway. The Navy ramp is much more crowded than in 1947, as VX-3, VC-4, and VC-33 had become large resident squadrons. Also note that the municipal airport terminal and two-lane entrance road, in the lower left of the photograph, have been completed. In the right photograph, the extension in length and addition of parallel taxiways to runways 4/22 and 13/31, as well as the increase of runoff areas for the other runways, to meet the demands of high-performance jets, are evident. (Above, NA; right, TH.)

NAVY V. MOTHER NATURE, ROUND THREE. With the importance of the all-weather, day/night role of the resident VC-4 and VC-33 composite squadrons, operations could not be halted by a south New Jersey snowfall, such as that of February 1955 shown here. In the photograph above, a lineman starts clearing snow from VX-3 F9F-6 Cougars. Below, a treaded tractor pulls a VC-4 F2H-4 Banshee across the snowy ramp. Although the nearby Atlantic Ocean tended to offer some moderating influence on winter weather, NASAC did suffer through periods of snow, ice, fog, and rain, making operations, especially at night, that more difficult and dangerous. A number of VC-4 and VC-33 airmen were lost in operations around NASAC. (Both, NA.)

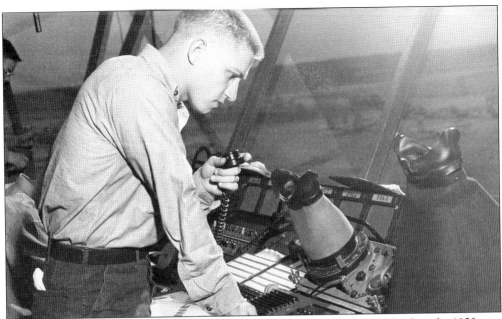

NASAC TOWER CONTROLLERS, 1954. While the level of operations at NASAC in the 1950s was not the same as when it was a fighter training base during World War II, the presence of three rather large squadrons (VX-3, VC-4, and VC-33) meant that the volume of air traffic was still significant. The original control tower was mounted atop the Operations Building. The hooded display is for the radar scanning the sky around the air station. The rectangular slips of paper located on the slanted panel in front of the controllers represent individual aircraft, arranged in the order of their arrival/departure from the field, and containing information about the individual flight plans. The Navy built a new, taller control tower, but it was not put into service before NASAC was closed in 1958. (Both, NA.)

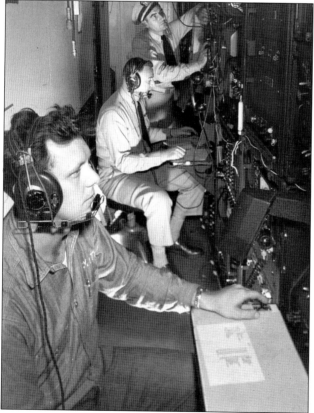

NEW GCA RADAR UNIT, JULY 1954. Dating back to World War II days, NASAC played a prominent role in developing radar-controlled flight. In the 1950s, NASAC was home of Navy GCA Unit 20 (GCA-20). Using a precision radar system, an approaching aircraft's vertical and horizontal position in relation to the proper flight path could be monitored, and corrections verbally provided to the pilot allowed safe landings on the darkest, rainiest nights. The photograph above shows a mobile GCA Unit that could be installed at virtually any airfield. From the left are the control truck, power supply, and radar/communications trailer. The left photograph shows the interior of the control truck, with two controllers and a supervising officer. Today, the same kind of alignment information is provided to the pilot electronically. (Both, NA.)

CORSAIR MAINTENANCE, SEPTEMBER 1952. Postwar, the Navy suffered from a high turnover rate, and training was a constant challenge. Here, a VC-4 chief instructs a group of line mechanics on the finer points of maintaining the 2,300-horsepower, Pratt & Whitney R-2800 Double Wasp engine of a squadron F4U-5N. This complex engine also powered the wartime Thunderbolt and Hellcat fighters as well as postwar Douglas and Lockheed airliners. (NA.)

SCHOOL DAYS, NASAC STYLE, 1952. The training of new maintenance personnel was made more challenging as the Navy transitioned to jet-powered aircraft. In many ways, a turbine engine is simpler, more reliable, and easier to maintain than a piston engine, but in 1952, the systems concept was quite new. Here, a chief goes through the key features of a Westinghouse J-34 turbojet engine, two of which powered the Banshee. (NA.)

CROMMELIN FIELD DEDICATED, MAY 1949. The Crommelin name is prominent in Navy history. Rear Adm. John G. Crommelin, right, was one of five Annapolis graduate brothers. Two were naval aviators who trained at NASAC but were killed in action. In 1949, an athletic field was dedicated in their honor; located off the end of runway 4, it still exists as part of the Air National Guard base. (NA.)

OPEN HOUSE, MAY 1949. The relationship between NASAC and Atlantic City was always a cordial one. The Navy welcomed the public to the air station every May for Armed Forces Day. This photograph shows a group of young men ogling one of the Navy's first operational jet fighters, the McDonnell FH-1 Phantom, with Corsairs and Avengers in the background. (NA.)

THE NAVY PARADES FOR MISS AMERICA, 1949. The Miss America Pageant began in Atlantic City in 1921 and was held there every September until 2004. This annual event was perhaps one of the most important for Atlantic City, and the Navy often participated. This photograph shows a Navy unit marching down the famous Boardwalk, passing the current location of the Trump's Taj Mahal Casino Resort. (NA.)

OPEN HOUSE, 1950. The 1950 Armed Forces Day open house gave the Navy the opportunity to show off its latest jet fighter, the F2H-2 Banshee of VX-3 (note the "XC" tail code). In the background can be seen a row of VC-4 SNB trainers, a single C-47 transport, and rows of midnight-blue Corsairs and Skyraiders. (NA.)

HELICOPTERS STILL A NOVELTY. The 1950 open house also featured demonstration of a Sikorsky HO3S-1 helicopter of Helicopter Squadron 2 (HU-2). Helicopters were still pretty new for most people, but this type would play a vital role rescuing downed Navy fliers from the sea and from behind enemy lines during the Korean War, which coincidentally would begin a month after this air show. (NA.)

ARMED FORCES DAY ON BOARDWALK. In addition to the usual open house, the 1951 Armed Forces Day was celebrated by parades and displays on the Boardwalk. In this photograph, a Navy Honor Guard follows a group of female Marines. Note the advertisement for a Nash Airflyte automobile. The Ritz Carlton Hotel in the distance still exists as condos but is now overshadowed by the Tropicana Casino Hotel. (NA.)

NAVY JETS ON PARK PLACE. For the 1951 Armed Forces Day celebration, the Navy brought some of its latest aircraft about 10 miles by road from the NASAC to Atlantic City for display in front of the Claridge Hotel at Park Place (the prestigious Monopoly board game address). The aircraft on the left is a VX-3 F9F-5 Panther, while on the right is a VC-4 F2H-2N Night Banshee. (NA.)

OPEN HOUSE, 1953. During the 1953 NASAC open house, Boy Scout Donald Brown got the thrill of his lifetime when he got to sit in a Marine Night-Fighting Squadron 114 (VMF[N]-114) F4U-5N Corsair. The Marine night-fighting Corsair was visiting from Marine Corps Air Station Cherry Point, North Carolina. (NA.)

MISS AMERICA PARADE, 1952. The close relationship between the Navy and Atlantic City is demonstrated again in this photograph, where sailors are escorting a float with Miss Naval Air Station Atlantic City as part of the Miss America Pageant Parade. The large building in the background was called Conventional Hall but is now called Boardwalk Hall. It still is the venue for concerts, sports, and other events. (NA.)

F2H-2P PHOTO BANSHEE ON DISPLAY. During the 1954 open house, NASAC's VX-3 showed off its photographic reconnaissance version of the Banshee twin jet fighter. McDonnell Aircraft built 58 of this version with an elongated nose to accommodate six high-resolution cameras, shown in front of the Banshee, which could take photographs simultaneously in the vertical and horizontal directions. (NA.)

IKE DEPARTING NASAC. President Eisenhower was a well-loved and respected president; local newspapers reported that Atlantic City welcomed his October 1953 visit with cheering crowds and waving flags. At the end of his visit, returning to NASAC, he took the opportunity, shown in the right photograph, to meet Navy personnel, including Lieutenant Harris, commanding officer of the WAVES (acronym for "Women Accepted for Volunteer Emergency Service") unit, more properly the Naval (Women's) Reserve. Captain Young, commanding officer of NASAC, looks on. In the photograph below, Ike gives his traditional goodbye wave from the entrance of Lockheed VC-121A *Columbine I* (Air Force 48-614), his first of three Constellation transports named after the official flower of Mamie Eisenhower's home state of Colorado. This aircraft is currently preserved at the Pima Air and Space Museum in Tucson, Arizona. VX-3 Banshees are behind. (Both, NA.)

VICE PRESIDENTIAL VISITS. Atlantic City has had its share of presidential visits with every sitting president—from Teddy Roosevelt to Jimmy Carter—flying in for an occasion. Vice presidents were also welcomed. In the above photograph, dated May 1950, Vice Pres. Alben Barkley with his wife, Jane, are greeted by NASAC commanding officer Johnson. They have just stepped off a Boeing B-17, converted to an executive transport, to speak at the Steel Workers Union convention. In the photograph below, Vice Pres. Richard Nixon is greeted by the local Atlantic City chapter of the Kiwanis Club in 1953. The international volunteer service organization had its conventions at Atlantic City in 1947 and 1953. Note that Vice President Nixon used a lowly U-4B Aero Commander twin-engine executive aircraft, one of two in the White House fleet. Travel at that time must have been a lot less comfortable than what current government officials enjoy. (Above, NA; below, ACFPL.)

ROYAL NAVY ADMIRAL VISITS. In December 1953, Royal Navy vice admiral John Hughes-Hallett visited NASAC to learn about Navy advances being made in night and all-weather fighter and attack capabilities. In the right photograph, Vice Admiral Hughes-Hallett steps from a Beech SNB onto the NASAC ramp. The photograph below shows him (center, smoking) attending a briefing with VC-4 and VC-33 personnel. Hughes-Hallett played a vital role in the wartime European campaign, working with Lord Mountbatten on planning the cross-channel raids and commanding the ill-fated Dieppe Raid. Most importantly, he was instrumental in devising the Mulberry artificial harbors used so successfully during the Normandy invasion. At the time of this visit to NASAC, he was the flag officer of the *Home Fleet* and later commanded the British carrier *Illustrious*. (Both, NA.)

VICE ADMIRAL HOSKINS INSPECTING NASAC. In December 1953, shortly after his promotion to Commander Fleet Air, Quonset Point, was announced, Vice Adm. John Hoskins inspected NASAC—one of the air stations under his new command. A 1921 Annapolis graduate, Hoskins gained his Wings of Gold in 1937. Assigned to the light carrier *Princeton* as prospective commanding officer, he lost his right foot during the Japanese air attack that resulted in its sinking. Despite this injury, he remained on active duty and commanded the new carrier *Princeton* as well as Korean War veteran *Valley Forge*. He also was the first Navy pilot to demonstrate the takeoff and landing of a jet aircraft on an aircraft carrier, when many "experts" said it was impossible. (Both, NA.)

HOLLYWOOD AND ATLANTIC CITY. The Hollywood production of *Caine Mutiny* was released in 1954, and the Navy played a role in promoting this movie starring Humphrey Bogart, José Ferrer, Van Johnson, Fred McMurray, Robert Francis, and May Wynn. In return, some of the stars, including Robert Francis and May Wynn (both pictured, with an unidentified Navy recruiting officer), participated in Navy recruitment tours, one of which included a stop at NASAC, as shown in these July 1954 photographs. In 1955, avid pilot Robert Francis and two passengers were killed when an aircraft he was piloting crashed in Burbank, California. Model and aspiring actress Donna Lee Hickey's big break was gaining the role of May Wynn in the *Caine Mutiny*; thrilled with this achievement, she changed her name to May Wynn! Subsequently, she starred in a number of television series in the 1950s. (Both, NA.)

ARTHUR GODFREY TAKING A JET RIDE, OCTOBER 1953. Known as the "Old Redhead," Arthur Godfrey was a noted radio and television broadcaster of the 1950s and early 1960s. He was a Navy veteran and lifelong aviation enthusiast and private pilot. In return for time spent in Navy public relations activities, he gained honorary naval aviator status and a ride in the backseat of a Lockheed TV-2 jet trainer at NASAC. (NA.)

CANADIANS DROPPING IN FOR A VISIT. In September 1953, due to rough weather off the Eastern Seaboard, the Royal Canadian Navy carrier *Magnificent* diverted her embarked air group to NASAC for a visit. This photograph shows Hawker Sea Fury piston engine fighter-bombers and VC-33 Skyraiders. During the Korean War, Lt. Peter "Hoagy" Carmichael shot down a Communist MiG-15 jet fighter while flying a Sea Fury from HMS *Ocean*. (NA.)

Four
GOODBYE NAVY . . . HELLO FAA

AIRWAYS MODERNIZATION BOARD, JULY 1, 1958. A 1956 midair airliner collision over the Grand Canyon spurred Congress to create the Airways Modernization Board (AMB) to improve airway safety. Its need for a field-testing experimental facility coincided with the Navy's departure from Atlantic City. Gen. Elwood Quesada received "ownership" of the former NASAC; he became the head of the Federal Aviation Agency (FAA), which replaced the AMB six weeks later. (GEH.)

PANORAMA, 1958. This photograph shows what the FAA inherited from the Navy. Known as the National Aviation Facilities Experimental Center (NAFEC), the 2,500-acre original Navy site, expanded to 5,000 acres, contained about 180 structures as well as long runways, radar equipment, and infrastructure. Early FAA experimental laboratories were located in uncomfortable converted wartime warehouses and buildings that had been built as "temporary" structures. (Stan Ciurczak.)

PANORAMA, C. 1968. In the austere budget climate, the FAA struggled adapting the existing Navy buildings for many years and instead invested in test equipment and facilities. One of the first new structures was the large aircraft hangar, shown under construction to the right of the new airline terminal site. Air National Guard F-100Cs line the ramp next to their new hangar in the left of the photograph. (Stan Ciurczak.)

AIR TRAFFIC CONTROL R&D AT NAFEC. By the end of the 1960s, it was clear that the rise in airline traffic would prompt a continued rise in the demand of air traffic control services in the decades to come. Research and development activities utilizing the FAA's simulation laboratory, shown in these photographs, contributed to the modernization of the ATC system funded by the Airport and Airway Development Act of 1970. In the photograph above, IBM engineers (the white shirts are a giveaway) work alongside FAA researchers in 1969 setting up the computer system. In the photograph below, NAFEC director Buck Commander (far right) checks out the ATC simulation laboratory equipment in 1973. (Both, Stan Ciurczak.)

AIR TRAFFIC CONTROL SIMULATION LABORATORY. The country's first laboratory simulating air traffic on radarscopes was used to develop the computer technology needed to assist air traffic controllers. Starting in the 1960s, the FAA began development of a system under which flights in certain "positive control" areas were required to carry a transponder radar beacon. This allowed for the identification and tracking of aircraft. Pilots in this airspace were also required to fly on instruments regardless of the weather and to remain in contact with controllers. The 1971 photograph above shows the extent of the simulator laboratory, where the entire National Airspace System could be reproduced. In the 1975 photograph below, controllers track flights from multiple positions during a simulation of traffic in the nation's air space. (Both, Stan Ciurczak.)

AIR FORCE C-5As TESTING ILS, 1973. In addition to the focus on air traffic control, NAFEC studied a wide variety of air safety challenges, including navigation, airport, and aircraft safety. The Instrument Landing System (ILS) is a ground-based system that provides pilots with information on the landing aircraft's position along the glide slope (vertical) and localizer (lateral) position, an electronic version of the earlier voice based Ground-Controlled Approach concept. (Stan Ciurczak.)

ONE OF NAFEC'S TEST FLEET. Over the years, NAFEC and its successor, FAA's William J. Hughes Technical Center, have operated a varied fleet of test aircraft. This one, a Douglas C-54 Skymaster transport aircraft, was used for airborne electronic tests during the late 1960s and 1970s. Depending on the specific testing needs and program duration, FAA aircraft flown from Atlantic City were either acquired or borrowed. (Stan Ciurczak.)

NEW CONSTRUCTION CONTRACT SIGNING. The Atlantic County Improvement Authority (ACIA) arranged Wall Street funding to construct new technical and administrative headquarters, allowing a move out of the old Navy barracks. From the left, Congressman (and facility namesake) William Hughes, FAA technical center director Robert Faith, and ACIA chairmen Albert Marks and Seymour Kravitz sign the contract for the new facilities. (Stan Ciurczak.)

GROUND BREAKING, TECHNICAL AND ADMINISTRATIVE HEADQUARTERS. From its establishment in 1958, NAFEC's estimated 2,000 employees largely worked in deteriorating buildings left over from the Navy. On September 20, 1978, Pres. Jimmy Carter, behind the podium in photograph, praised the center's development of many new technologies for enhancing safety in flight. Smiling New Jersey congressman Peter Rodino, of President Nixon's impeachment hearing fame, is standing in the dark suit on the right. (Stan Ciurczak.)

DEDICATION OF NEW HEADQUARTERS. On May 29, 1980, Vice Pres. Walter Mondale (center, behind podium), New Jersey governor Brendan Byrne, FAA administrator Langhorne Bond, technical center director Joseph Del Balzo, and the first man to break the sound barrier, Gen. Chuck Yeager (sixth from right), were present to dedicate Building 300, the new FAA William J. Hughes Technical Center's Technical and Administration Headquarters Building. (Stan Ciurczak.)

SATELLITE VIEW OF NAFEC. This satellite imagery shows the William J. Hughes Technical Center around 1998, after the new headquarters were built and the Coast Guard Air Station, just to the left of the FAA hangar, was established. The new terminal, ANG base, and old terminal site are also evident. Only two (runways 13/31 and 4/22) of the Navy's four runways remain, although traces of the other runways are still visible. (Stan Ciurczak.)

NATIONAL AIRPORT PAVEMENT TESTING FACILITY (NAPTF). Recognizing that existing airport pavements might not accommodate newer, heavier aircraft, the FAA established an industry and government-working group to determine the full-scale testing needs. The result was an FAA/Boeing jointly built and operated facility. Completed in April 1999, NAPTF is the world's largest enclosed full-scale test facility dedicated to airport pavement research. The building shown in these photographs houses a test runway 900 feet long and 60 feet wide. A total of 20 test wheels can be configured to represent two complete aircraft landing gear trucks under a varying set of conditions, landing weights, and speeds to accelerate pavement damage. Test results allow the FAA to set design standards and ensure compatibility with new generations of aircraft. (Both, Stan Ciurczak.)

RUNWAY OVERRUN PROTECTION. Developed at the William J. Hughes Technical Center, Engineered Materials Arresting Systems (EMAS) are extensions to existing runways constructed of high-energy absorbing materials of specific strength. They are designed to crush under the weight of commercial airplanes as they exert deceleration forces on the landing gear. EMAS is particularly useful for airports where the land at the end of a runway is unavailable. A standard EMAS installation extends 600 feet from the end of the runway. In the photograph above, an FAA Boeing 727 plows through an EMAS test pad at Atlantic City. The photograph below shows the EMAS installation at the end of Boston's Logan Airport runway 15R/33L, designed to prevent aircraft from overrunning into Boston Harbor. Since 1999, fifty-nine systems have been installed, and seven potentially disastrous runway overrun accidents have been prevented. (Both, Stan Ciurczak.)

FAA Control Towers, Atlantic City. The original Navy control tower, sitting atop the Operations Building, can be seen in earlier photographs in this book. In the mid-1950s, the Navy built a more modern control tower, but NASAC was disestablished before it could be put into use. In February 1959, that tower (shown as it is today in the photograph at left) was moved a half mile across a runway to a location just south of the then-existing Municipal Airport terminal. A new 229-foot control tower (below) was dedicated in 1988. Its location is about 600 feet west of the old tower, almost a mile northwest of the new airline terminal building. (Left, Author; below, Stan Ciurczak.)

WILLIAM J. HUGHES TECHNICAL CENTER TODAY. One of the FAA's current aircraft is the Bombardier Global 500, shown overflying the technical center. The roof-mounted control tower allows the simulation of tower procedures while the actual airport tower is top left. The pond was originally installed to provide cooling for the computer systems but is no longer needed, Canada Geese now use the pond. The FAA hangar and Coast Guard Air Station are to the right. (Stan Ciurczak.)

ALLEGHENY COMMUTER DHC-6 OVER BOARDWALK, C. 1970. Despite the fact that Bader Field (seen under the aircraft's nose) was the first commercial "air port" with airline and sightseeing service starting in the 1910s, Atlantic City has had difficulty sustaining airline service. Even with construction of a new Atlantic City International Airport terminal at the FAA's William J. Hughes Technical Center, airline service has been sporadic. (ACFPL.)

"NEW" MUNICIPAL AIRPORT, c. 1955. Even from the first concept prior to the takeover by the Navy, a municipal airport terminal was planned for the western edge of the site, off Tilton Road. Although the access road, shown in this photograph, was eventually paved by the Navy, the terminal building was not constructed until the mid-1950s. Today, the terminal is gone, replaced by the airport fire department. (ACFPL.)

MUNICIPAL AIRPORT TERMINAL. The original terminal was built while the Navy was still operating NASAC. The very modest building shown here, with a total area of roughly 3,000 square feet, served as the terminal until the new terminal was built in the mid-1960s, about three quarters of a mile south of this location. (ACFPL.)

PLANS FOR NEW TERMINAL, 1960. Recognizing that the existing municipal terminal at the former NASAC site was inadequate to attract airline service, the FAA's NAFEC and Atlantic City awarded contracts for the design of a new terminal building. One of the designs, shown here, contained many features that were ultimately built. The terminal's first phase of construction, completed in late 1962, was 300 feet long. (ACFPL.)

ATLANTIC CITY INTERNATIONAL AIRPORT. Since the first phase of construction was completed late in 1962, the terminal has undergone a number of upgrades and expansions. This photograph shows how it looks today in the midst of yet another expansion. With the recently announced termination of service by AirTran, only Spirit Airlines currently serves Atlantic City. What will it take to induce other airlines to join Spirit? (Author.)

ATLANTIC CITY MURAL. One of the nicest features of the new terminal, shown as it appears today, is the Atlantic City Boardwalk mosaic mural. Composed of 55,000 pieces, it shows the city's pre-casino skyline at the time of construction in 1960, with all the great hotels depicted. Today's skyline is quite different, with only a few of the earlier hotels still in existence. (Author.)

SPIRIT AIRBUS A319. Spirit Airlines is the largest, ultra–low cost airline network serving the United States, Latin America, and Caribbean. It is equipped exclusively with modern Airbus A319, A320, and A321 jetliners. They serve six cities with nonstops from Atlantic City (KACY), with connections to dozens more. The A319-132 *Spirit of Fort Lauderdale* shown here, about to return to KACY from Tampa, is powered by two IAE V2524-A5 turbofan engines. (Author.)

Five

GUARDING THE STATE, PROTECTING THE NATION

NJANG F-84F THUNDERSTREAKS. On August 5, 1958, the New Jersey Air National Guard's 119th Fighter Squadron moved from Newark Airport to NAFEC, as part of the Tactical Air Command. In this 1962 photograph, Thunderstreaks, flown by Capt. Thomas McGivney, Maj. Charles Young, and 1st Lts. William Charney and Mike Piecenck, return to Atlantic City from a mission over Warren Grove Bombing Range. (NJSA.)

JERSEY DEVILS' THUNDERSTREAK. Prior to its move from overcrowded Newark Airport in 1958, the squadron (known as the Jersey Devils, after the legendary creature said to inhabit the Pine Barrens) had just transitioned from North American F-86E Sabres to Republic F-84F Thunderstreak fighter bombers. Put onto active duty in October 1961 as part of the Berlin crisis call-up, the unit remained at Atlantic City until released back to state control in August 1962. (NJSA.)

F-86H SABRE, C. 1963. On October 15, 1962, the unit reached group status with federal recognition of the 177th Tactical Fighter Group. They also transitioned to the ultimate version of the Korean War vintage Sabre, the F-86H. North American Aviation built almost 500 of this improved model, known as the Hog, which incorporated a low-altitude bombing system, nuclear capability, and the replacement of six machine guns with four more-potent cannon. (NJANG.)

F-100C Super Sabre, c. 1966. In September 1965, the 177th Tactical Fighter Group transitioned to the supersonic single-seat F-100C and two-seat F-100F Super Sabre fighter bombers. In the photograph above, G-suit wearing pilots are "stepping" to their aircraft, passing in front of one of the unit's two-seaters. These garments, more properly called anti-G suits, are designed to prevent a loss of consciousness under high acceleration maneuvers. In the right photograph, weapons specialists are loading 2.75-inch folding-fin aerial rockets into underwing pods in preparation for a sortie to nearby Warren Grove Bombing Range. This bombing range, which contains strafing lanes, simulated ground targets (tanks, parked aircraft, and antiaircraft sites), and even "villages," was inherited from the Navy and is still in use. (Above, NJANG; right, NJSA.)

LAST CHANCE CHECKS AT ATLANTIC CITY. It is common practice for tactical aircraft to pause at the departure end of the runway for last minute checks for hydraulic or fuel leaks, loose access doors, remaining safety pins, and when appropriate, arm munitions. This photograph shows 119th TFS/177th TFG Super Sabres undergoing such safety checks prior to departing on a training mission. (NJSA.)

ARMED FORCES DAY, C. 1965. Established by President Truman in 1947, it is celebrated on the third Saturday of May. In this photograph, Atlantic City's 119th TFS/177th TFG personnel march down the Boardwalk past the Steel Pier as part of the annual Armed Forces Day celebration. The Steel Pier has been an Atlantic City landmark since 1898, surviving storms, fires and economic downturns. (NJSA.)

PRESENTATION OF THE COLORS, C. 1968. On January 26, 1968, the unit was called to active duty for the Pueblo crisis. The unit, sporting the tail code "XA," was transferred to the 113th TFW at Myrtle Beach AFB. In this photograph, the unit's personnel are on review in front of their F-100s, now painted in their Southeast Asia camouflage scheme. (NJSA.)

ENTER THE "THUD." In June 1969, the unit returned to Atlantic City and state control. A year later, the 177th Tactical Fighter Group transitioned aircraft again, this time to the famed F-105B Thunderchief, known affectionately as the "Thud." This base entrance sign from 1972, adorned with models of the F-105, declared the group's role in "serving the state and the nation." (NJANG.)

LINE MAINTENANCE OF F-105s, c. 1971. This Mach 2–capable aircraft was originally designed as a nuclear bomber, but in the Air Force and Air National Guard, its role was conventional bombing and interdiction strikes. The New Jersey Air National Guard flew the F-105B from both Atlantic City and McGuire Air Force Base. A number of the New Jersey Thunderchiefs were former Thunderbird display team aircraft. (NJANG.)

NJANG THUNDERCHIEF, c. 1971. This F-105B is taxiing out for takeoff carrying a practice bomb dispenser on its belly pylon. Flying over heavily defended targets in North Vietnam, losses of F-105s were so high they had to be withdrawn from frontline service—the first and only time this has happened to the USAF—more due to flawed micromanagement of the war by Washington than failings of the Thud. (Author.)

THE DELTA DART IN NEW JERSEY. In 1972, the unit converted to the F-106A/B Delta Dart. Its designation was changed to the 119th Fighter Interceptor Squadron, 177th Fighter Interceptor Group when it was reassigned to the Air Defense Command. Here, four squadron T-33 trainers are parked next to F-106s. One of the former Navy hangars is being demolished in the background. (NJANG.)

F-106 WITH NEW HANGAR. Capable of Mach 2.8 and altitudes of nearly 60,000 feet, the F-106 Delta Dart is considered by many to have been the Ultimate Interceptor. It was armed with a M-61 six-barrel Vulcan cannon and four AIM-4 Falcon missiles. It became the standard air defense fighter in the Air Force and Air National Guard. In the 1960s, a new hangar replaced a Navy hangar located at roughly the same spot. (Patrick McGee.)

F-106B Two-Seater. Convair built 277 single-seat F-106As and 63 two-seat B versions. Here, a two-seater plows through the humid skies near Atlantic City. The "hump" on the spine of the aircraft is the air-to-air refueling door. All Delta Dart units had a few two-seaters assigned for crew proficiency training. (NJANG.)

Bear Hunting, Jersey Style. As part of their Air Defense Command tasking, the 177th FIG maintained an around-the-clock alert, guarding the coast from Long Island to the Virginia Capes. During the height of the Cold War, it was not uncommon to come across long-range Russian aircraft, such as this Tu-95 (NATO code name Bear) turbo-prop bomber, transiting to Cuba. Note the Bear's tail gun at a raised angle, a "nonhostile" signal. (NJANG.)

BEAUTIFUL PORTRAITS OF A BEAUTIFUL AIRCRAFT. In the photograph above, a pair of 177th FIG Delta Darts is flying off the coastline of New Jersey. In the photograph below, a four-ship, or echelon formation of four Delta Darts, peels away from the camera. There is a saying in aviation that "the nicer a plane looks, the better it flies." From that perspective, the F-106A Delta Dart was definitely a good flying airplane. It served on active duty with the US Air Force for 28 years, which was longer than most of its contemporaries. The New Jersey Air National Guard was the last unit to fly the F-106; on August 1, 1988, the last three aircraft were flown to the Arizona "Boneyard" for conversion into target drones. Not just a few tears were shed at that event. (Both, NJANG.)

VIPER OVER THE BOARDWALK, JANUARY 2003. During 1988, the unit transitioned to the F-16A/B Fighting Falcon (more commonly, "Viper") retaining its air defense role. In 1992, the unit designation was changed to the 177th Fighter Wing/119th Fighter Squadron (tail code "AC"), transitioning to the F-16C/D in 1994, becoming a general-purpose unit. Here, a 119th FS Viper flies past the Boardwalk, Pier Shops at Caesars, Caesars Palace, Trump Plaza, and the ubiquitous Boardwalk Center. (NJANG.)

VIPER FOUR-SHIP OVER LBI. This January 2003 photograph shows four New Jersey Air National Guard 177th FW F-16Cs banking to the south over Long Beach Island (known locally as "LBI"), taking the scenic route back to their Atlantic City base from Warren Grove Bombing Range. LBI's permanent population numbers less than 10,000, but during the summer it swells to more than 100,000 vacationers. (NJANG.)

VIPER TWO-SHIP, FEBRUARY 2009. The 177th Fighter Wing, one of the 20 remaining Air National Guard units flying the Viper, now flies the F-16C Block-30 version. Generally, the Air National Guard's F-16s are the oldest in the Air Force inventory, but thanks to the high level of pilot experience and dedicated maintenance, these units are still very potent forces. (NJANG.)

JERSEY DEVIL OVER IRAQ, JULY 2004. From October 2001, the unit supported Operations Noble Eagle, Southern Watch, Joint Guardian, Jump Start, Enduring Freedom, and Iraqi Freedom. Wing personnel have deployed to Iraq, Afghanistan, Qatar, United Arab Emirates, Oman, Djibouti, Yemen, Uzbekistan, Kyrgyz Republic, Turkey, Albania, England, and Germany. It was the first unit (active duty, Guard or Reserve) to fly more than 1,000 Operation Noble Eagle missions. (NJANG.)

177TH FW F-16, ATLANTIC CITY INTERNATIONAL AIRPORT. In this October 2001 photograph, a Jersey Devil Viper is shown banking over its home base. The semicircular outline of the old municipal airport is now the home of the 177th FW alert sheds and airport fire station. The dark paved runway is 13/31, while the light colored one is 4/22. The outline of Navy runway 17/35 is still visible. (NJANG.)

WARREN GROVE RANGE. The 177th FW operates Detachment 1 at the Warren Grove Bombing Range, located about 19 miles northeast of its base. The 9,500-acre site, nestled in the New Jersey Pine Barrens, is used for air-to-ground practice by Air National Guard, Navy, Marine, and Army units. This 175th FW, Maryland ANG, A-10 has just made a strafing pass with its 30-millimeter Avenger Gatling cannon. (Author.)

HISTORIC AIR STATIONS MERGED. In the mid-1990s, the US Coast Guard merged two of the oldest and most historic air stations—Cape May, New Jersey (above, showing a Sikorsky HH-3F Pelican), and Brooklyn, New York (below). Air Station Cape May, commissioned in 1926, was the first permanent air station. It was used for search and rescue as well as Prohibition-era rumrunner patrols off the New Jersey seacoast. Air Station Brooklyn at Floyd Bennett Field was established in 1938 as an air rescue seaplane base and later became the birthplace of rescue helicopter operations. Air Station Atlantic City was commissioned on May 18, 1998, at the FAA's William J. Hughes Technical Center. (Above, NA; below, Author.)

USCG, ATLANTIC CITY. Air Station Atlantic City is housed in a 63,000-square-foot facility. The L-shaped floor plan provides for operations and maintenance control located between "ready" and maintenance hangars. Collapsible doors enclose each of the hangars' 140-foot openings. Here, a newly delivered MH-65D sits on the ramp, with the ready hangar to the right and the maintenance hangar behind the tail. (Author.)

RESCUE SWIMMER, ATLANTIC CITY. Air Station Atlantic City keeps two MH-65D Dolphin helicopters on 30-minute response status. Once airborne, they can speed to a rescue site at more than 145 knots. The crew includes two pilots, a flight mechanic, and a rescue swimmer. This photograph shows a rescue swimmer jumping from the chopper during a demonstration off the Atlantic City beach. (Author.)

"DON'T FOOL WITH MOTHER NATURE." Although Atlantic City does enjoy some really nice weather, as this book has chronicled, it also has suffered from sand storms, hurricanes, snowfalls, and fog—conditions that aviators must be wary of. Taken from USCG Dolphin helicopter, this photograph shows thick coastal fog enveloping Atlantic City's Resorts, Taj Mahal, and Showboat hotel and casinos; the famous Steel Pier juts into the ocean. (USCG.)

DOLPHIN AND "OLE BARNEY." USCG MH-65Ds fly daily missions from Atlantic City, about 40-percent for proficiency training with the rest routine patrols of ports and waterways. Here, a Dolphin passes Barnegat Inlet and Barnegat Lighthouse, perhaps, the most widely recognized icon of the Jersey Shore. In 1609, Henry Hudson came across this inlet, which he named Barende-gat, meaning "an inlet with breakers." Across the bay is the town of Barnegat. (USCG.)

Manhattan before 9/11. Air Station Atlantic City mission supports search and rescue, maritime law enforcement, port security, aids to navigation, and environmental protection support for an area including the coastlines and waterways stretching from Connecticut to Virginia. This photograph shows a HH-65C Dolphin helicopter from Atlantic City exercising with a Coast Guard cutter just south of the World Trade Center. (USCG.)

Detachment MH-65D Passing the Washington Monument. Of the current inventory of ten MH-65D helicopters, eight are stationed at Atlantic City. Since the 9/11 terror attacks, Air Station Atlantic City has also provided a 24 hours a day, seven days a week, two helicopter alert detachment at Reagan National Airport, Washington, DC, under the operational control of the North American Aerospace Defense Command (NORAD). Here, one of its Dolphins passes the Washington Monument. (USCG.)

Six

THE SPIRIT OF ATLANTIC CITY

THE SPIRIT OF ATLANTIC CITY, 1943. The generous and patriotic people of Atlantic City contributed to a War Bond campaign, raising $76,552 to purchase a P-47D Thunderbolt from Republic Aviation. This aircraft was assigned to Capt. Walker "Bud" Mahurin of the 56th Fighter Group/63rd Fighter Squadron stationed at Halesworth, England. (MAAM.)

Captain Mahurin at Briefing. When the *Spirit* was assigned to Michigan native Bud Mahurin, he was asked if he wanted to change the name. He replied that since it was a gift of the people of Atlantic City, he wanted to keep that name. Here, Mahurin (left) is reviewing the route of his squadron's mission escorting heavy bombers on a raid over Berlin. (MAAM.)

Congratulations after Brunswick Raid, 1944. In this photograph, 1st Lt. Robert S. Johnson (left) congratulates Captain Mahurin in front of his *Spirit of Atlantic City* for becoming the highest-ranking ace in the European theater of operations. Note the crew chief adding a "kill" marking to the *Spirit's* scoreboard. Mahurin is credited with 21 European kills before being shot down, 19 of which were flying the *Spirit*. (NA.)

MAHURIN IN KOREA, 1952. After being shot down over France in 1944, Mahurin evaded capture and joined the 3rd Air Commando Group in the Philippines after a War Bond tour of the United States. He shot down a Japanese bomber in the Philippines but was downed himself. He commanded the 4th Fighter Group in the Korean War, where he is credited with 2.5 MiGs but was shot down and suffered 16 months of torture in North Korean captivity. (MAAM.)

ROBERT G. DOSE. Navy captain Dose was the commanding officer of NASAC's Air Development Squadron 3 (VX-3) from June 1955 to June 1957. He arrived with an impressive service record. In this 1944 photograph, squadron commander Dose (left) is pictured with Air Group commander Joseph Clifton, with squadron commanders Vince Hathorn and Bill Rowbothan, on *Saratoga*. (NNAM.)

VX-3 FJ-3 Furies. On August 22, 1955, Capt. Robert Dose conducted the first operational test of the Mirror Landing System (MLS), flying a VX-3 Fury, like the VX-3 aircraft shown near NASAC, to a successful trap on the carrier *Bennington* steaming in the Atlantic Ocean. Another squadron pilot flew a Grumman F9F Cougar for the first nighttime MLS landing. (TH.)

First Crusader Ejection. Captain Dose has the dubious distinction of being the first fleet pilot to eject from a Crusader. While flying over New Jersey, he tangled with a flight of Pennsylvania Air National Guard F-94s, but his engine flamed out. To avoid crashing in a populated area he glided to the ocean before ejecting. He returned to NASAC after being rescued by a Navy Albatross amphibian, as shown here. (GEH.)

FIRST CARRIER-TO-CARRIER TRANSCONTINENTAL FLIGHT. VX-3 Commander Dose was tasked to demonstrate the new Crusader on a historic nonstop flight from Pacific to Atlantic Ocean carriers. On June 6, 1957, he and wingman Lt. Comdr. Paul Miller flew from *Bon Homme Richard*, 50 miles west of San Diego, to *Saratoga*, steaming east of Jacksonville. The flight required refueling their F8Us over Dallas from AJ-2 Savage tankers, as shown here. (NNAM.)

DOSE ABOUT TO TRAP ON *SARATOGA*. During the flight, the two Crusaders held Mach 1.7 as they climbed to the altitude of 43,000 feet. Approaching Jacksonville, they had to skirt an enormous thunderstorm. Diving down to the carrier, Dose passed the deck at more than 450 knots, pitched out pulling seven-Gs, and managed a perfect "three-wire" landing, shown in this photograph. (Vought.)

DOSE AND MILLER GREETED BY IKE. The transcontinental flight commemorated the 13th D-Day invasion anniversary, and appropriately, President Eisenhower greeted the aviators on *Saratoga*. The flight took about 3.5 hours, a record that still stands. Not enjoying the bustle of photographers and journalists, Dose discretely had their aircraft positioned on the catapults allowing them to make a quick exit back home to Atlantic City. (Both, NNAM.)

ENSIGN JESSE BROWN. Fighting Squadron 32 (VF-32) was one of the first squadrons established at NASAC in 1943. During the Korean War, VF-32 was on *Letye* alongside NASAC's VC-4 and VC-33. VF-32's squadron included Ensign Jesse Brown (shown here), the first African American naval aviator, and his friend Lt. Thomas Hudner. (NA.)

MEDAL OF HONOR RECIPIENT HUDNER. On December 4, 1950, Ensign Brown and wingman Lieutenant Hudner were flying their F4U-4 Corsairs near the Chosin Reservoir, North Korea, when Brown's aircraft was shot down. Looking down at the wreckage while circling above, Hudner saw Brown trapped in the cockpit and bravely crash-landed his own Corsair nearby to try to aid his squadron mate. Unfortunately, despite his repeated attempts to free Brown, he was unsuccessful and Brown died in his crushed cockpit. On April 13, 1951, Hudner received the Medal of Honor, the first of the conflict, from President Truman (above) and the grateful thanks from Ensign Brown's widow, Daisy (below). After a tour as an instructor, Hudner was assigned to VX-3, where he tested the latest Navy jet aircraft in the skies above Atlantic City. (Both, NH.)

RISKING HIMSELF TO SAVE OTHERS. On August 31, 2000, then-major David Haar (above) and three 177th FW squadron-mates departed for training sortie over the Atlantic. About 35 miles out, the engine of Haar's F-16 (similar to the photograph below) threw a turbine blade, severing oil lines. He immediately turned back to base. Approaching the airport in marginal weather with a crippled jet, he noted the crowded beaches, roads, and populated casinos, and selflessly decided to turn back toward the sea to avoid any chance of his jet falling into a populated area. He waited until the last possible moment as he finally crossed the coast at Brigantine, ejecting at 1,700 feet, well below minimum recommended altitude. He was rescued by a state police boat with minor injuries. Major Haar received the Air Force's prestigious Air Medal for his heroic actions. Lieutenant Colonel Haar (ret.) now flies for Spirit Airlines out of … Atlantic City. (NJANG.)

"Let's Roll" Nose Art Dedicated. In this photograph, Lisa Beamer, the widow of *Flight 93* hero Todd Beamer, and 177th FW's Col. Mike Cosby unveil a F-16C with *"Let's Roll,"* Todd's last words when he led a group of passengers in an attempt to regain control of the hijacked airliner. The motto *"Spirit of 9-11"* superimposed on a sword, an American eagle and a US flag are also part of the nose art. (NJANG.)

Navy Memorial. Near the NJANG base entrance is a simple memorial dedicated to the members of Composite Squadron 4 (VC-4) and the memory of their fallen shipmates. More than 40 squadron members made the ultimate sacrifice during the 1948–1958 period they flew from NASAC. This memorial also can serve as a reminder of the countless Navy, Air National Guard, Coast Guard, and FAA aviators who served their nation from this historic airfield. (Author.)

Discover Thousands of Local History Books Featuring Millions of Vintage Images

Arcadia Publishing, the leading local history publisher in the United States, is committed to making history accessible and meaningful through publishing books that celebrate and preserve the heritage of America's people and places.

Find more books like this at
www.arcadiapublishing.com

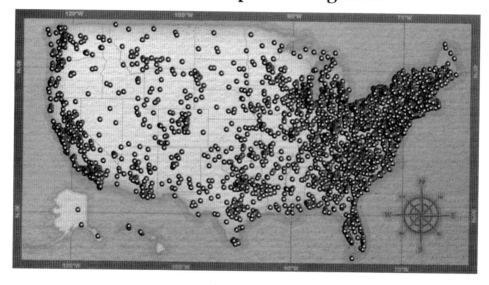

Search for your hometown history, your old stomping grounds, and even your favorite sports team.

Consistent with our mission to preserve history on a local level, this book was printed in South Carolina on American-made paper and manufactured entirely in the United States. Products carrying the accredited Forest Stewardship Council (FSC) label are printed on 100 percent FSC-certified paper.